The Dynamics of Property Location

Why is property located where it is, and how has this process changed in recent years? A number of factors, such as social change and technological development, have affected location and these are considered. Value, the way changing patterns of location are measured, is examined and there is a discussion of rent contours. The book considers location in the retail industry, looking at the theory, hierarchy, clustering and dispersal. The move to out-of-town sites, with its three waves of decentralisation, is described. Central place theory, dating from the 1930s, is discounted as being obsolete and misleading. Finally the book covers offices, industrial and residential property.

Key features:

- Clear and concise
- Strong practical basis
- Focus on retail sector
- Rich in valuation data

Russell Schiller, PhD, FRICS spent nearly 30 years developing property research at CB Hillier Parker, becoming a partner and Head of Research. In 1984 he was made Honorary Professor of Land Economy at the University of Aberdeen. The book sets out the text in a simple and non-technical manner, imbued with a strong practical sense, to provide a solid textbook for the Land Economy or Land Management undergraduate student and junior professional.

The Dynamics of Property Location

Value and the factors which drive the location of shops, offices and other land uses

Russell Schiller

London and New York

First published 2001 by Spon Press
11 New Fetter Lane, London EC4P 4EE

Simultaneously published in the USA and Canada
by Spon Press
29 West 35th Street, New York, NY 10001

Spon Press is an imprint of the Taylor & Francis Group

© 2001 Spon Press

Typeset in Sabon by Steven Gardiner Ltd
Printed and bound in Great Britain by
TJ International Ltd, Padstow, Cornwall

British Library Cataloguing in Publication Data
A catalogue record for this book is available from the British Library

Library of Congress Cataloging in Publication Data
Schiller, Russell.
The dynamics of property location / Russell Schiller.
    p.   cm.
1. Real property – Valuation. 2. Store location planning.
3. Offices – Location – Planning. 4. Homesites – Planning.
5. Land use – Effect of technological innovations on. I. Title.
HD1387 .S297 2001
333.33'2 – dc21   2001023408

ISBN 0-415-24645-8 (hbk)
ISBN 0-415-24646-6 (pbk)

To Daphne, who else?

Contents

Figures

Tables

Acknowledgements

I would like to thank CB Hillier Parker and my former colleagues there for their help and for letting me use much of the material used in this book, in particular Jo Semple Piggott who prepared many of the charts. Thanks are also due to Angus McIntosh for his help with the Internet chapter, and to Frances Charlesworth on whose expert typing I depended on heavily.

This book was written during a long illness and I would like to acknowledge the encouragement I received from Murray Rayner, Harold Couch and Michael Hallett.

1 Introduction

Preface

Imagine a committee of booksellers meeting to decide where to open a centre to sell second-hand books. The idea is to bring together collections of books on such a scale that they serve the whole country and the centre acts as a point of reference for all book buyers. The committee would no doubt agree that such a centre should be located where it is easy to reach. It seems so obvious that it is likely that little time would be wasted on this point.

In fact just such a centre has opened, not as it happens as the result of a committee meeting but naturally, following the success of a few enthusiasts. The place is Hay-on-Wye, which must rank as one of the least accessible places in the country, lying as it does on the Welsh border to the west of Hereford. Its population is tiny, barely enough to qualify it as a town at all, yet its literary festival and its huge range of second-hand book shops has brought it national fame.

After the First World War what was claimed to be the biggest office block in the world opened. In answer to a Trivial Pursuit-type question as to where it was, it is reasonable to suppose that many people would opt for the centre of New York, or the centre of one of the great European or American cities. If they did they would be wrong because the new building in question was the new headquarters of General Motors, built not in the centre of Detroit but far out beyond the suburbs in what could justly be claimed as the first example of office decentralisation.

Both these stories suggest that the study of location can hold surprises and that common sense may not be the best guide. In neither case does the location actually chosen fit the theory as it is commonly taught by the main academic disciplines which cover this subject: economics, geography and land economy.

In explaining GM's decision, Joel Garreau (1991) wrote:

> GM distanced its headquarters from downtown in a place grandly called New Centre. . . . It was accessible to downtown – by car. But

it was also close to GM's factories – some of which were purposely located outside the city and its taxing power. It was very close to swanky neighbourhoods like the Boston-Edison area where the executives lived. And it offered far more land than the old downtown both for expansion – and parking.

The revolution caused by the switch to dependence on the car is one of the themes running through this book. Indeed GM's decision must be one of the first examples of this on any scale.

The lesson of Hay-on-Wye is also one of car dependence. Although miles from the nearest motorway, Hay is still within half a day's drive from London and most of the rest of the country. Although less accessible than most parts of Britain, Hay is still sufficiently easy to reach by car for it to function adequately. The fact that it takes, say, four hours to reach it rather than three is not of overriding importance to the visitor once the decision to make a major trip has been made.

Hay, therefore, functions as a destination, a magnet which attracts people from far and wide, and because of that there are many other locations where it would function equally well. Costs of land and property there are among the lowest in the country and there are also plenty of redundant shops and commercial buildings for the second-hand book traders to occupy. Ironically it is the economic failure of Hay to thrive as a market town which has made it attractive to the low rent paying second-hand book trade. For its present function Hay works well and its location is adequate but there are other places which might have functioned just as well. Having been established by its founders, it has the size and momentum to continue and to attract visitors as a destination.

Dynamics

The aim of this book is to throw light on how location works in practice and to see if there are any patterns which can be found which can explain them. It deals mainly with commercial property, that is to say shops and offices, although it also covers industry and housing. In particular it examines the changes which have occurred during the last third of the recent century.

This period, 1965–99, has been a time of great change in location, a time during which many of the old ideas have been swept away. A student studying in the 1960s would have been taught theories developed before the Second World War covering industrial locations and the hierarchy of settlement patterns dealing with shops and offices. These were based on assumptions which did not always apply in practice, but the underlying idea was that the forces they described were permanent. They were not dynamic.

Now the basic premises themselves are questioned and shown to be of little relevance. In their place we find a dynamic world of continuous change which makes it difficult to replace old theories with new. This book offers no new theories but concentrates instead on identifying the forces which influence location and describing how they work on the ground.

Economic man may still exist, minimising cost and maximising utility, but he gives only part of the picture. He certainly would not have chosen Hay-on-Wye. It is not so much that this approach is wrong as inadequate; there is a great deal more to the equation than that. Indeed one of the themes which emerges throughout the book is the importance of an attractive environment such as Hay. Increasingly, when there is a considerable choice where location could be successfully sited, as in Hay, it is perceived attractiveness which is the vital factor.

The key factor and the most dynamic of all in its effect is the role of the car. The car transforms many of the old ideas of accessibility, in fact it stands some of them on their heads. The town centre, for instance, used to be the area of maximum accessibility because public transport routes radiated out from there. The coming of widespread car usage changed that completely around and because the car made it congested, the town centre became the least accessible of locations. People continued to use the town centre, but they now do so despite its location rather than because of it.

The car is a glutton for space. It demands road space while being driven, parking space during the day and, often, its own garage at night. It is the enemy of the type of high density living that about 90 per cent of the population experience, leading to ever rising levels of congestion. Nevertheless it is an enormous liberator, enabling people to drive to Hay-on-Wye for a day's book browsing, for example, something that would otherwise be impossible.

The arrival of the car as the majority mode of transport enormously increases the mobility of ordinary people. It enables them to work in a much wider range of jobs, to have a greater choice of shopping, and to use a wider range of leisure facilities. In terms of location for all the various uses that need access to the public (employment, schools, medical facilities, shops, services, leisure, etc.) it has the effect, too, of increasing choice. These uses are freed from the constraint of town centre and public transport routes and can serve their clientele from a greater variety of sites.

The car in fact is the victim of its own success. Congestion goes a long way to nullify its advantages, leaving a blank prospect of roads clogged both for private cars and public transport, as well as other essential vehicles. Nevertheless it seems difficult to imagine the old pre-car location pattern returning. The toothpaste is out of the tube.

Forces

There are two fundamental forces at work, the desire to cluster and the desire to disperse. The whole central part of the book, chapters 5 to 8, is devoted to considering how they interact in terms of retail location.

The desire to cluster leads to high density development and increases the user's ability to choose between different goods and services. Even where shopping itself has moved away from the town centre it often prefers to locate within shopping centres. This happens mainly with shops selling clothing and other durable goods. Other types of shops, such as supermarkets often prefer to disperse.

The desire to disperse leads to low density development, which is often nearer where people actually live than the town centre itself. It results, almost inevitably, in the increased use of the car. Despite the efforts of the planning system to contain activity within urban areas, population and employment have tended to disperse more widely, too, so that the move of commercial activity to disperse has merely been in line with this trend.

Shops and services which serve a low density population have the problem that to be near their customers they need to operate on a small scale, but to run a successful business they need to be big. Most choose the latter and ask their customers to use the car. This results in out-of-town shops being bigger than those in town centres, sometimes much bigger.

America is often a good laboratory to observe locational forces at work. With abundant space, it is a country with a planning system far weaker than Britain's. There can be seen greater extremes of clustering and dispersal than exist in Europe. Housing is commonly at a very low level of density, particularly in the newer mid-western cities such as Phoenix and Denver. At the same time, America is the home of the skyscraper, which rise in city centre clumps even in the newer cities just mentioned. Perhaps the existence of the opposite forces of clustering and dispersal in extreme forms even in a near-open market planning system shows that they are indeed the two key locational forces at work.

Elements

Dividing a subject such as location into a series of chapters is difficult because the elements or dimensions organise themselves in different ways. Some overlap and others suggest different ways of organising the material. In the main the chapters here refer to different land uses such as shops, offices, industrial and housing, but the first five chapters deal with elements which apply to all these various land uses.

It is worth touching on three of these elements – density, value and access. They are all key factors in determining how the two forces mentioned above affect what actually appears on the ground.

Density is expressed in terms of building height and number of floors (plot ratio) and the proportion of the site which is covered by building (site cover). Although some land uses such as shops and offices tend to be associated with high levels of density there is no use which cannot function perfectly well at relatively low density. American experience suggests that even where the building density itself remains relatively high, as with regional shopping

Table 1.1 Competing land values

	Price per acre of a small well-located out-of-town site, May 1986		
	Retail*	Industrial	Residential
Croydon	£1,000,000	£375,000	£350,000
Basingstoke	£700,000	£250,000	£350,000
Birmingham	£600,000	£85,000	£100,000
Manchester	£375,000	£70,000	£60,000

Source: CB Hillier Parker

Note
*The price for superstore sites can be much higher.

centres or corporate office headquarters, the parking provision and land-scaping surrounding it means that the site cover stays low.

The trend has been for density to fall with increasing car use. The accessibility given by the car has put in being a movement for all human activity to be spread evenly across the land rather than be clumped together in towns. Although there are still the remains of the rural-to-urban movement which has been in existence in the developed world for centuries, there is also a new stronger trend towards dispersal which is continuing despite the best efforts of the planning system to stop it (see chapter 11). It is of fundamental importance.

The trend to low density is not one recognised by many who describe the future. There seems to be a widespread view that future living will be modelled on Gotham City with people living in towers under domes which provide clean air. In fact this comic strip vision of the future is based on the Manhattan of the 1930s. People who have the choice have since then mostly voted with their wallets to live the majority of their lives at low density.

Value, the second element, reflects the interactions of the locational forces at work. It is a measurement of the strength of competing demands for land, and it deserves a chapter to itself (chapter 4). Value reflects the balance of supply and demand for space, particularly demand. It also shows the relative strength of competing land uses, as shown in Table 1.1.

Data and maps on value have been conspicuously lacking in most books on this subject, and this has partly been due to the fact that consistent time series of value have not been publicly available. The author spent much of his career at a major firm of chartered surveyors collecting and analysing value data, and it is hoped that one of the benefits of this book is to make this evidence available to a wider audience.

Access, the third of the elements, is as vital as any. If the location process is thought of as a machine, then value is the measure of temperature or performance, but access is the liquid which is pumped round, or the oil which makes it possible to function. As we saw with the two examples earlier, access is the first question which is normally asked when considering location.

Again we return to the car. It is because the car has transformed accessibility that the other changes have followed. The increase in accessibility which it has brought about has made location a less precise study than it was before. Land users have become footloose. They have been given a wide choice of where they can site their operations, and this in turn leads to the final two points of this introduction.

Themes

Two themes resulting from increased accessibility are worth emphasising. The first of these is the growing importance of the attractiveness of the environment in location. The footloose occupier of property, liberated by widespread car use, increasingly places attractiveness high in the selection criteria when choosing where to locate. It is a theme which is referred to in several chapters since it applies widely to the different types of commercial and residential property.

In shopping we see very high rents being asked and paid in heritage towns such as Bath, Chester and York. Waterfronts have been converted to shopping centres in Liverpool, Hull, Southampton and several other towns where the attractiveness of a developed waterfrontage is felt to be an important consideration in attracting shoppers; the environment and the specialist shops themselves combining to make the trip enjoyable.

In offices there is the example of the growth of Bournemouth which attracted a large number of financial service firms when they decided to move out of London. A substantial number of cities have seen their central areas expand with new office development in the direction of the attractive area occupied by wealthy housing. This trend can be found in London, Paris, New York, Barcelona, Leeds and Birmingham and many other cities. Garreau noted that General Motors built its new headquarters near the 'swanky' suburbs where the executives lived.

In industry there is a general move from rust belt to sun belt, with factories, population and values all rising in the southern American states at the expense of the older industrial areas in the north. A warmer climate is held to be among the major causes. In France there is also a sun belt effect with above average growth along the Mediterranean coast. Research centres are particularly free to locate where they want and it is therefore no surprise that Sophia Antipolis, the major French centre for hi-tech research, should be sited in the hills behind Cannes and Antibes. There are many other examples which could be quoted.

The second theme is the trend towards fewer but bigger operations. Here again one is spoilt for choice for examples to illustrate the point. Thus there are fewer shops but more retail floor space, fewer garages but more petrol sold. Almost any public sector use that can be named has tended to close the old and the small and replace them with a lesser number of big and modern operations. This can be seen in prisons, libraries, local authority offices, schools, colleges, hospitals and law courts.

Many of these uses were previously in town centres, along with private sector activities such as breweries and the relocation to bigger premises invariably led to a move out from the town centre. It is only fair to point out that the move to out-of-town shopping, widely deplored as harmful to the town centre, is no more than following the example of the public sector.

The town centre has suffered because it has been unable to accommodate this change. It simply does not have the space, and as a result has lost its monopoly of what used to be considered central place activities. The pressure for large new operations had been building during the 1960s after three decades of recession, war and austerity, but it was growing car use which allowed them to take place.

Once again we are back to the car, and the changes it has wrought since the 1960s. In many ways this book is the story of what those changes have been. It is an exciting story and it is not over yet.

2　Location and politics

Politics

Development is a political act, there is no getting away from it. It is tempting when looking at different types of location and levels of values to seek patterns and laws as one might do in biology. Certain land uses such as banking and fashion retailing do tend to concentrate on high-value land in city centres, but there is no simple determinist rule that says that this must always be the case.

Location decisions are made by fallible people who often fail to optimise the things which economics says they should (access to markets and labour, for example). This is particularly the case during times of rapid change.

But more important than this is that the occupier of property, the retailer looking for a shop, the company looking for a head office, the importer seeking a distribution warehouse, and the house builder wanting to buy suitable land, is only one actor on the political stage where land-use decisions are made.

Development occurs as the result of the interaction of political forces, and this interaction has the result of reducing the choice available to the occupier. There are a number of political forces at work seeking to influence the use to which land is put, and the developer and occupier are by no means the most powerful of these forces.

Three examples will make the point. The volume house builders tried hard in the 1980s and 1990s to promote concentrations of new houses in planned villages in the south-east only to be turned down again and again. Second, of development anywhere in the county-wide London have proved largely futile for several concerns the tightening up of controls during tail development which has virtually ended the g centres such as Bluewater Park in Kent, and

in the field could give many more examples of point is not whether these restrictions are right nly arguments in favour of the three restrictions described. The point is that the result limits the choice available to the occupier and often creates a stark dilemma: if the location and type of building that is wanted is not available, the choice is either to do nothing or to accept second best.

This dilemma is sharply felt by foreigners, typically Americans, seeking to open chains of branches in Britain. They arrive in a chartered surveyor's office and put a list of locational requirements on the table. In practice these are often very similar. They want a flat, rectangular site with a main road frontage offering good visibility and ample parking and they want the site to be within ten to twenty minutes' drive of an adequate catchment of their target market.

There are increasingly sophisticated techniques using GIS which can produce sites which best serve the target market. What there are not, and what the Americans find very difficult to swallow, are available sites in these places which would be allowed planning permission. Having planned to open say 40 branches in a year, the newcomer often finds output of two or three a year is all that is achievable in practice.

The result of this is that for many occupiers the whole question of location theory, if not irrelevant, is in practice less important than knowing what is acceptable and what is available. The mundane, but necessary, takes precedence over the more esoteric.

In studying location, therefore, and in observing how different occupiers cluster together or disperse themselves across the land, it is vital to understand that they are where they are, not as the result of scientific laws but because of a rather crude interplay of political forces. Locational forces are certainly at work but they can only be seen through the dark glass of politics.

The retailer – disturber of the peace

The actors in the development process can be conveniently divided between those who instigate change and those who react to it. In retail, pressure for change comes mainly from the more modern half of retailing, the national multiples. They are continually looking to modernise their form of operation and they are supported by the developers and financial institutions which service them.

The other actors in the drama are all reactive to a greater or lesser extent. They are happy with things as they are and wish to be left in peace. The starting point is therefore to examine this pressure.

Retailers want more space. They want modern facilities, good access, and other things too, but it is the desire for more space which, like ice in a crevice, has cracked open the pattern of retailing which applied roughly from the First World War to the late 1960s.

Retailers are far from being alone in this. The taste for more elbow room has been widespread among a whole range of activities, as we saw in chapter 1. There are fewer shops than there were. The number has halved from over half a million to a quarter of a million since 1950. Retail floor space in contrast continues to rise in total in line with retail sales.

The reason for the need for more space is not only that it offers greater choice but also that it facilitates modern operations. It makes possible economies of scale sufficient to justify spacious modern machinery and higher standards of comfort for staff. It also provides room for large modern delivery lorries to service the shop at the back, out of the way of the shopper.

Retailers wanting more space have a choice. They can either try and stay in the traditional location in the town centre or they can move out of town. Either way they create a problem. If they stay their expansion, which is often accommodated in a new shopping centre produced by a comprehensive redevelopment, may threaten the historic form of the city. If the redevelopment itself escapes that criticism, then the extra lorries coming to supply the store and the extra car parking demand are likely to add pressure to the city centre. If retailers decentralise then they are open to the charge of draining life from the city centre, using valuable green field land and of being inaccessible to those without cars.

Retailers answer these criticisms by claiming with some justice to understand the shopper better than the other actors in the retail planning process. They have to be intimately attuned to the shoppers' wishes in order to survive. They are scornful of the others, the planner, the politician and the conservationist who have probably never tried to bring home the family groceries towing a child on a cold, wet winter's day without the use of a car.

Retailers are fond of saying that the shopper votes for them with her purse at their modern, computerised check out, but this is the essence of their problem. This support is inarticulate. The same people who throng the air-conditioned shopping centre, the superstore or retail warehouse are likely to sign petitions calling for the preservation of the corner shop and deploring the intrusion of soulless shopping complexes. The cash register vote the retailer gets is impressive, but it is only ballot box votes which count.

The planner – guardian of the faith

The actors who have to cope with the retailer's pressure are a varied group. They include the planner, the conservationist, the politician and small retailers, but of these the planner is the leader, the professional. It is one of the tragedies of the current debate that the profession of the land became split between planners and chartered surveyors as it began to develop in the middle of this century.

Today the two groups often co-operate, but there still tends to be a deep difference in attitude between them. Planners nearly all work in the public sector for local authorities and they see their jobs as reconciling various non-commercial objectives. The retailer's desire to maximise profit therefore cuts little ice.

The chartered surveyor, on the other hand, works mainly in the private sector servicing private sector clients and marches to a business drum. There is a tendency to regard the planning system as a barrier stopping industry and commerce giving the public what they want. Not surprisingly planners tend to the left and surveyors to the right in politics.

The planning system is a powerful one, enabling a proposed development to be judged for its aesthetics, its affect on roads, parking and other infra-structure, and its impact on existing shopping, among other things. But these powers are limited in practice because planners often find themselves out of date, running after events and using an obsolete plan.

Every decade produces new forms of commercial development, be it retail parks and factory outlet centres in the retail field, or office parks and high bay warehouses elsewhere. These new formats are often of a different scale to what they replace, as was explained earlier, and they make planning demands on the area where they locate.

They need to be planned, but the process of making a plan takes so long that by the time the plan is approved it is obsolete and makes no mention of the new land uses currently being demanded. The effect is to weaken the planning process. Sometimes a developer successfully appeals to central government, and the local authority reluctantly finds it has a development it did not want imposed on it. Sometimes the local authority gives consent contrary to government guidelines because it wants the employment the new development offers. Neither outcome is as desirable as if the proposal had been properly thought through and integrated into the plan.

It is often said in criticism that the British system of government is over centralised, and that too much power is held by central government and not enough by regions and local authorities. At first sight the strength of the planning laws seem to support this. The DETR (Department of the Environment, Transport and the Regions), for example, has wide powers to 'call in' planning decisions, thereby taking them out of the local authority's hands.

Despite this, it is noticeable that Britain has found it harder to crack down on out-of-town development than most continental countries. The DETR has followed an increasingly tough anti-out-of-town line throughout the 1990s, yet retail parks and factory outlet centres, virtually all of which are out-of-town, continue to receive consent. As Table 2.1 shows, if local authorities want to give consent for out-of-town development, it seems that it is difficult for central government to stop them. It is only with big developments of say 500,000 ft² (50,000 m²) or above that government action appears to be completely effective.

Table 2.1 Retail parks in the pipeline (000 m²)*

Figures at June of each year	Under construction	With planning consent	Proposed
1990	650	1,860	2,430
1991	320	1,640	2,120
1992	130	1,370	1,110
1993	330	1,460	680
1994	420	1,510	1,010
1995	200	1,780	1,350
1996	310	1,500	1,180
1997	220	1,530	1,060
1998	290	1,480	1,170
1999	210	1,190	810

Source: CB Hillier Parker

Note
*Construction activity shows little fall 1990–1999.

Riding the hierarchy

Planners have all been taught central place theory and the functions of the retail hierarchy at college, and consider it a valuable conceptual framework for judging shopping development. The idea of retail hierarchy and its implications for modern location planning are discussed in later chapters (5 and 6). The concept of hierarchy is introduced here because of its effect on planners' thinking and how it affects modern development.

Most of the structure plans of the 1980s and some later plans make preserving the retail hierarchy a central policy objective. In practice this means allocating a rank to each town centre in the plan area and defending it with policies such as encouraging new development in the town centre and stopping change of use from retail to services in the main retail frontage.

Applications for free-standing superstores and retail warehouses are difficult to fit into the hierarchy idea. They are too big to fit into the neighbourhood centres which form the base of the hierarchy, indeed they are not centres in the traditional sense of the word at all. They also threaten to take trade away from the very town centre which the policy is designed to protect.

Throughout the 1970s and much of the 1980s, the argument that a free-standing superstore might benefit shoppers by offering greater convenience and choice, and might also benefit the town centre by removing pressure which conflicted with durable shopping, gradually won the day. But for some planners it was a chain of reasoning which was irrelevant because it ignored the concept of hierarchy.

Not surprisingly, retailers found a way of using the hierarchy to their own ends. The first example of this was the building of modern district centres. The traditional ranks of the retail hierarchy are set out and explained in the master plans of the post-war new towns (Stevenage is a good example). They run from the single corner shop, through neighbourhood centres (perhaps five to 15 shops in size), and district centres (big enough for the main weekly food shop). Then on to town centres which offer clothes and other durable goods, to regional centres with department stores and other specialist shops.

The district centre was the most vulnerable to the impact of the superstore. Many district centres in the 1970s were strung out along radial roads leading out from big towns, and suffered from heavy traffic, narrow over-crowded pavements, and a stock of obsolete shop property. They were considered unsatisfactory by retailers, planners and shoppers alike.

If the free-standing superstore could add a few shops and possibly non-commercial functions too, like a clinic, sports centre or library, it could call itself a new district centre. At once it would fit into the hierarchy and become something the planner could understand. In the early years of superstore development, say 1970–85, retailers and developers found it paid them to describe their proposals as a new district centre.

The second example of how the idea of hierarchy became twisted is rather more subtle. Instead of moving outside the hierarchy by developing on a virgin site, the retailer can sometimes gain the advantages of decentralisation by moving down the hierarchy from a larger to a smaller centre. Because the European settlement pattern is so ancient, the ideal decentralised location on the edge of the city may, for the very reason it is ideal, have already become the centre of a smaller community some hundreds of years earlier. The decentralising retailer can therefore claim to be reinforcing the hierarchy by boosting the smaller centre even though he is at the same time denuding the larger centre.

A good example of this phenomenon is Sutton Coldfield, on the edge of Birmingham where the Gracechurch Centre of 500,000 ft^2 (46,000 m^2) was developed in what had previously been a modest town centre. Another example is Redditch to the south of the Black Country. Here the Kingfisher Centre of 770,000 ft^2 (72,000 m^2) was grafted on to the existing meagre town centre, completely transforming the shopping provision in the town centre.

In each of these cases a traditional small town centre accommodated what was in effect a Brent Cross-type regional centre attracting trade from a long way beyond the original catchment area, and competing with Birmingham and other major centres in the vicinity. In terms of impact on the top centres of the West Midlands hierarchy, the effect of these two centres was surely no less than if they had been free-standing, but there were accepted as conforming with planning orthodoxy because they were within two, albeit small, existing town centres. Thus they were deemed to reinforce, not weaken, the retail hierarchy.

Passing mention should be made here of Merry Hill, the giant 1,450,000 ft^2 (135,000 m^2) shopping development near Dudley. This received planning permission in the late 1980s when the government was experimenting with planning-free Enterprise Zones, and adopting a free market approach to land use. The only way it could be argued that it fitted into the hierarchy was as an additional high order centre, an argument which cut little ice with most of the West Midland planners at the time.

The two examples given, new district centres and major shopping developments in small towns on the edge of conurbations, show how the idea of hierarchy adapted to new forms of retailing and to pressure for decentralisation. It would be wrong to suggest that this resulted from cynical manipulation by developers and retailers. The development process resembles a juggernaut and is too slow and public for anything so machiavellian to occur. Rather, a consensus emerged from discussions which both sides found satisfying, though for quite different reasons. In fact, in spite of the radically different objectives pursued by the actors in the political process of development, there has been a surprising amount of consensus in recent decades, the subject of the next section.

The search for consensus

So far in this chapter there has been an emphasis on the conflict between commercial forces and the other actors, the planners, conservationists, politicians and small retailers. It was necessary to present them in the traditional British adversarial way to highlight the issues involved. In many respects, however, there has been remarkable agreement. The majority of retailers and developers, even those promoting out-of-town schemes, continue to join with the other interests in wanting to preserve healthy retailing in the town centre. In this section we look at the big picture to see how far there is a consensus and when that consensus changed. Throughout the first three-quarters of the twentieth century there was wide agreement that the town centre was the right place for modern shops and offices and that it should contain as much as could be supported by the market (argued by the retailers) or justified by central place theory (argued by the planner). Heroic efforts were made in the 1960s and early 1970s to accommodate modern commercial forces, and after three decades of depression, war and austerity these forces had built up some strength.

This was the era of the inner ring road, many of which were planned and some actually completed. Plans were made to accommodate rising car usage, pavements were narrowed and multi-storey car parks built. It was the time of the first major wave of comprehensive shopping developments running from the mid-1960s to the late 1970s, and it was the time of a boom in office building when many towns acquired steel and glass and mini-skyscrapers which have never been superseded.

Looking back from the end of the century it is easy to criticise much of this investment for the destruction of historic buildings and much-loved medieval streets which it caused. We can see now the futility of trying to provide unrestricted car access, indeed with the wisdom of hindsight many would agree that the redevelopment exuberance of the 1960s would not be appropriate today.

The point is that at the time there was a consensus that modern commerce should be encouraged, not thwarted, and that the town centre was the right place for this to happen. Investors, retailers, planners and politicians were in agreement, and the voice of the conservationists who objected on a few occasions was a whisper compared with what it later became and was muted precisely because the consensus was so widespread.

The first crack occurred in food shopping, which we can now see as the first of three waves of decentralisation, and as so often, it was technology driven. Self-service and the supermarket trolley led to a rapid increase in store size and the corresponding weight and volume of goods carried out by the shopper. This change is described in chapter 8, but the significance of it here is that it caused the supermarket operators to challenge the conventional wisdom that the town centre was the best place for the new, larger supermarkets.

In the town centre the shopper would emerge onto the pavement staggering under the weight of heavy bags to be jostled on a crowded pavement, perhaps recently narrowed as part of a new traffic plan. Supermarkets realised that the problem of the extra weight and bulk of shopping, and conflict with other town centre users could be eased by allowing the trolley to be wheeled straight to the boot of the car, and for that a large out-of-centre site was needed.

Asda pioneered free-standing superstores with car parking in their Yorkshire and Lancashire heartland, but in the south there was strong opposition to allowing food shopping to de-centralise. In the early 1970s J. Sainsbury called a conference at St John's College, Cambridge of the elite in the retail planning world and showed a film of town centre food shopping. Harassed mothers festooned with pushchairs and heavy shopping bags were seen struggling with steps and doors as they made their way to a distant car park or bus stop. The audience laughed but in time the point went home.

Throughout the 1970s a debate raged on the role of food shopping in town centres and it was won by the supermarkets. The supermarkets and their consultants used central place theory to argue that the weekly food shopping trip was different from the trip made to buy durables and that the congestion and litter caused by food shopping actually impeded the more leisurely comparison shopping.

Certainly for large centres with the majority of their shops selling clothing and other durables, a convincing case was made that the loss of mainstream bulk food shopping would do no harm and might even free up the centre to function more effectively in its role as a comparison centre. It was only in the 1990s when supermarkets began to open on the edge of smaller district centres and small country towns that town centre impact became a serious issue.

Up till then it had been possible to argue that the first wave of decentralisation had had little adverse impact on town centres, but the position with smaller centres was different. Suburban centres and small market towns function as district centres (see chapter 6), and the new supermarket competed directly with that function. The policy commonly adopted to deal with this problem has been to encourage the supermarket to locate within comfortable walking distance of the town centre so that it complements rather than competes with it.

Incidentally it is worth noting that the role of the supermarket in larger shopping centres is still debated internationally. In agreeing to the separation of convenience (i.e. food) shopping for durable shopping, Britain decided to follow the American model. But food shopping is considered an essential anchor to even the biggest centres in much of the rest of Europe, particularly southern Europe.

The coming of the third wave

By 1980 opposition in principle to free-standing out-of-town superstores was largely over and a period of rapid growth followed. The number of superstores increased from 172 in 1980 to 776 in 1992. This was followed in the 1980s by a second wave of decentralisation involving retail warehouses. There was little opposition in principle to these and they followed in the wake of superstores using similar arguments.

The driving force behind the expansion of retail warehouses was the revolution in DIY retailing. This was a new type of retailing, stimulated by the rising level of home ownership, which had never existed in the town centre. Its origins lay in builders' merchants' yards and so it was difficult to argue that it was bleeding the town centre.

Such an argument could not be used for furniture, carpet and electrical retailing which also moved rapidly out into retail warehouses and which formed part of a triangle with DIY of dominant retail warehouse operators. This type of retailer had been concentrated in town centres but badly needed space and it was obviously hard for the town centre to supply this on a scale which matched the 20,000–50,000 ft² (2,000–5,000 m²) offered by the out-of-town warehouses. True, furniture and electricals were 'durables' and therefore ought, according to central place theory, to locate within the hierarchy in town centres, but little attempt was made to keep them there on theoretical grounds. The argument to allow decentralisation of furniture and electricals was helped by the fact that retail warehouses and retail parks looked different from town centre retailing. They looked like industrial warehouses and were often located on what had previously been industrial land. There were no small shop units on retail parks to make them appear like normal out-of-town shopping centres.

The third wave involved clothing and other forms of comparison retailing, and just as the second wave had been facilitated by the success of the first, so too did the third wave follow closely on the arrival of the second.

It is possible to date the third wave to a single date: May 1984. It was then that Marks & Spencer announced it was to open out-of-town stores. In the annual report of that year the chairman blamed local authorities for failing to meet 'the requirement of the car-shopping public'. For Marks & Spencer, the country's leading clothing retailer and without doubt the bellwether of the comparison retailing flock, to make such an announcement was an end to consensus indeed.

Up to that point the consensus view was that decentralisation was under control and that clothing and comparison retailing, the backbone of the larger centres, was not under threat. Once Marks & Spencer had made their announcement the climate changed. A whole range of comparison retailers responded by announcing plans for out-of-town formats, and developers began to publicise out-of-town proposals. All told some 24 schemes of regional centre size (over 500,000 ft^2, 50,000 m^2) were announced by June 1986, and although the government quickly stopped the flood, the consensus was never to be the same again.

The planners fought hard to stop the third wave but with limited success. Large developments were 'called in' by central government and stopped, all except a handful which fell outside normal planning control because they were in Enterprise Zones, but others went ahead on retail parks.

Enterprise Zones were areas designated to be free of planning control and business rates to encourage investment. They were areas of such industrial dereliction that it was felt they would remain eyesores unless development was given a special stimulus. In fact they became the only place where large-scale retailing was allowed to take place. As a result three of the four regional out-of-town shopping centres (see chapter 6 for definition) that were built then were in Enterprise Zones. These were Metro Centre, near Newcastle (the first), Merry Hill in the Black Country, and Meadowhall near Sheffield. The fourth, Lakeside, near the M25 at Thurrock, was near an old gravel pit in an unattractive environment.

It is a paradox that retailing, that most location-sensitive of land uses, should end up making major investments in Enterprise Zones on sites so unappealing that major incentives were felt necessary to attract any type of investment at all. It is a paradox too, because environmental attractiveness is a major determinant in attracting investment. It runs as one of the threads running through this book. It was not an attractive environment, however, which determined the location of the big four out-of-town centres but a populous catchment area and the ability to get planning permission.

Planners were handicapped in opposing out-of-town development after 1984 by a more laissez faire central government policy, particularly during the time of Nicholas Ridley as Secretary of State in the late 1980s. This policy led to a major boom in retail development. Strangely, the laissez faire peak in Britain occurred at a time when France and most other continental countries were tightening controls against out-of-town development.

The ending of the consensus in favour of the town centre for comparison shopping in 1984 resulted in the re-emergence of the division between the thrusting retailer and developer on the one hand, and the other actors on the other. This was described earlier, only now it concerned fundamentals.

Retailers claimed they were reluctant to desert the town centre and Marks & Spencer continued to remind people that the overwhelming majority of their investment was in the town centre and they have been active in promoting local authority-run town centre management schemes.

Some blame the politicians for letting them down. John Lewis, which has several of its department stores out of town, argues that they have opened these stores only because the government has failed to protect their town centre investments. They joined with the developers Slough Estates to sue successfully Welwyn Hatfield Local Authority for breach of contract when consent was given to the out-of-town Hatfield Galleria development, thereby competing with their town centre department store in Welwyn Garden City.

Whilst planners and conservationists have consistently opposed out-of-town development, the commercial actors, investors, developers and retailers are in a dilemma. Market forces still favour out-of-town and they find them hard to oppose. At the same time the majority of their own investment is in town centres and they want to protect them as badly as the planners and the others.

Britain therefore seems poised, neither properly in nor out of town. It lies half way between the protected town centres of the Germans and the abandoned town centres of America. It is perhaps not a very comfortable position to be in. Strangely, the question which the conservationist should ask but does not do is whether the conservation of the town centre is better achieved by trying to accommodate modern commerce or by banishing it. That is a debate which has failed to take place and it lies outside the scope of this book. It might result in the re-emergence of a new consensus which could help solve the current commercial dilemma.

3   The importance of technology

The impact of technology

In the previous chapter we saw how location decisions were severely constrained by political conflict and the pressure of competing land uses. In this chapter we look at the effect of changes in technology on the demand for space. Technology is often overlooked as an influence in location theory, but its power can be more devastating than politics.

Fifty years ago, for example, railway marshalling yards where goods trains were shunted occupied large swathes of inner city land. Dockers and their problems were a major feature in labour relations and it was accepted that the docker with his hook, like Marlon Brando in *On The Water Front*, was the way goods were handled.

The revolution in goods handling, entirely driven by technological change, is discussed below, but it had the effect of making the whole vast area of London docks obsolete, stretching as it does for miles either side of the Thames from Tower Bridge to beyond Greenwich. Today, Docklands is known not for shipping at all, but as the home of the Canary Wharf skyscraper, and all London's imports and exports, bigger in volume than they were, are handled by a few container births around Tilbury.

Another example is the introduction of the steel girder into building around the fourth quarter of the nineteenth century. This happened first in New York, and enabled Manhattan and Chicago to sprout skyscrapers to heights unimagined before. Of course this would not have been workable without the parallel development of the modern lift, which Otis demonstrated at the New York Exposition of 1853. It was the twin novelties of the steel-framed buildings and fast electric lifts which transformed the skyline of American cities. These inventions may have been spurred on by economic and locational pressures, but without them no amount of pressure would have brought about the famous Manhattan skyline of today.

High-rise offices and the inventions which made them possible, permit huge increases in office density and make face-to-face communication easier. They allow vertical circulation to supplement the usual ground level horizontal circulation. They also allow a more efficient use of public transport, notably underground railways, and the provision of a wide range of business services close at hand. The skyscraper is not the universal solution to the need for high office density and some world cities, such as those in Europe, appear to function satisfactorily without them, but the steel girder and lift are now universal, and an indispensable part of any office complex almost irrespective of height.

Retailing has been affected, like the docks, by the distribution revolution, but there have also been changes in the way goods are merchandised which result from technological change. The supermarket would not work without the trolley and the modern shop window could not exist without plate glass. The improvement in glass technology has made possible the shop front with its display of goods which we take for granted today. In Jane Austen's *Pride and Prejudice*, written around 1800, Mrs Bennett advises her daughters on the 'warehouses' they should visit to buy their trousseaux. As the word suggests, they would have had little on display apart perhaps from a small-paned bow window to attract the passing pedestrian. The reason was not because of lack of merchandising skill in 1800, but simply because the glaziers of the day could not provide a large and safe enough area of glass to permit it.

As retailing has evolved it has exploited changing building technology. Modern shops benefit not only from large, tough, plate glass windows but from steel frames, lifts and escalators too. Indeed one of the reasons that shop sizes have been able to increase is because a steel girder above a plate glass window allows an old brick-built shop to have a wider frontage than was possible before.

From retail warehouses and superstores to shopping centres, modern retail development uses a scale and sophistication of construction which has steadily increased, often making obsolete developments as little as 20 or 30 years old. The control room of a modern shopping centre resembles the bridge of a ship, with screens showing closed-circuit TV pictures all over the complex and rows of dials monitoring the workings of the air-conditioning system which keep thousands of shoppers in comfort in a controlled space.

As with the skyscraper, technology has responded to pressure for change. Retailing has been moving to bigger shops and stores and bigger and more sophisticated shopping centres. The demand has steadily been for bigger floor plates uninterrupted by pillars. The bigger space permits the display of a wider range of goods to attract the shopper with opportunities for design which create an exciting atmosphere. It was the pressure to change which drove the technology, but without the breakthrough in technology no amount of pressure would have brought about the world we now see.

The car and its effect on behaviour

In purely locational terms nothing has had a greater effect than the car. It would be no exaggeration to say that this one invention threatens an urban tradition stretching back beyond classical times. Cities built before the car find great difficulty in accommodating it and those built since, like Phoenix in USA have a density so low they bear little resemblance to those older cities. The car transforms the mobility of the people who use it and, because of its demand for space, it also transforms the environment in which it operates. These two perspectives are looked at in turn, starting with the effect on the car user.

Access to a car enormously increases mobility, indeed it probably had a greater effect on human movement than any other single invention. This is apparent in commuting, shopping, and the pattern of holidays and social activities, all of which have been radically altered since the arrival of widespread car ownership. The car offers flexibility and control, the chance to travel at the precise time desired, and in door-to-door comfort with choice of music and protection from cold, wind and rain. It offers room for companions and space for luggage, and it is extremely difficult for public transport to compete.

The speed that the car travels increases the range of facilities which the car user has available, and, contrary to what might be expected, the effect is greatest over shorter distances. It fact it is for trips just beyond walking distance, say a quarter of a mile to three miles, where the difference between the car and other modes of transport is at its greatest, and these are the majority of all trips taken. For a car user, a journey of a mile, say to the supermarket, is simple, taking no more than five minutes. Without a car, such a journey becomes an excursion, needing planning if public transport is involved, and could easily take half an hour. If it is decided to walk, there is the problem of carrying shopping back on the return journey.

The increase in mobility offered by the car has a number of effects. First, it increases the total number of trips taken. People visit distant relatives more often or go for a drive in the country instead of staying at home. Second, it increases the choice available to the car user in terms of jobs, leisure and places to shop. By using the car to commute, the number of jobs available within practical reach of home is enormously increased. The Census of Population has included a question on where people work for some decades and the trend shows lengthening journeys to work in an increasingly dispersed pattern. The old pattern whereby people worked either locally or in the nearest big town has changed to a veritable spider's web of complex cross-country journeys now made possible by the car.

Table 3.1 shows the position according to the census in the St Albans district. Less than half of those living in the district also work there (1,527 out of 3,464). Most striking is the 21 per cent increase (1981–91) in commuting into the district. The census records that the work-flow across the district boundary is far from local. Only a third of those commuting out do so to the surrounding county. In fact the census lists 45 districts where the outflow is significant enough to record. Of course St Albans has an excellent train service and has long been a centre for rail commuting to

Table 3.1 Commuting in St Albans: 10 per cent sample of those economically active in the St Albans district

(a) 1991 Census

	Males
Living in the district	3464
Working in the district	2665
Living and working in the district	1527
Living in the district but working outside	1923
Working in the district but living outside	1136

(b) 1981–1991

	Males *1981*	*1991*	
Working in the district	2484	2665	+7.3%
Working in the district but living outside	938	1136	+21.1%

(c) Mode of travel in 1991 for all persons

	Living in the district but working outside	*Working in the district but living outside*
Train	644	63
Bus	38	54
Car as driver	1925	1505
Car as passenger	93	114
Total	**2803**	**1811**

Source: 1981 and 1991 Census, *Workplace and Transport to Work* tables

London, but this accounts for only 23 per cent of out-commuting. In fact the overwhelming majority of journeys to work are by car. Table 3.1 includes an extract from the census which shows that for the fast growing category of in-commuting, rail and bus are dwarfed by car use.

This increasingly defused pattern has in turn encouraged a more diverse pattern of employment. Firms can draw staff from a wider catchment area than before and so are free to choose more varied places in which to locate. In practice this may mean choosing a site which is cheaper and offers more space, and this works to the advantage of the small town or office park against the big centre. In particular it puts a premium on an attractive location such as a riverside or a small, pretty historic town. Not surprisingly the numbers working in small, attractive Thames-side towns like Henley and Marlow have grown disproportionately fast.

The effect of the car on retail location is even greater than the effect on employment. This is dealt with in later chapters, but in summary there are three main effects. First, the retail hierarchy which grew up to service shoppers travelling by foot and public transport became largely obsolete.

The car led to catchment areas overlapping and shoppers were no longer tied to the nearest shops. Second, it encouraged specialisation. Shops selling esoteric goods such as model railways or equipment for the disabled, which need a huge supporting catchment population, can thrive in fringe locations or small towns. According to central place theory they should be found only in large centres, but once their customers use the car then they are free to choose a cheaper location which is likely, as an added bonus, to have easier parking for their customers.

The third consequence is that greater mobility encourages retailers to increase the choice they offer by locating in larger stores. This benefits out-of-town at the expense of the town centre, because out-of-town offers space, low costs and customer parking to an extent which the town centre finds hard to match. It also benefits large towns which offer more choice at the expense of small towns. In summary, the car has meant that retail location in its entirety has had to be rethought from top to bottom.

The car and the environment

We have already seen how the car encourages the decentralisation of employment and that the effects on retailing are equally strong. Cars find it difficult to cope with high density land use. They function much more efficiently in areas of low density and they are therefore the enemy of the town centre. Before the arrival of widespread car use in the 1960s and 1970s, town centres thrived because of their accessibility. Main roads radiated out from them as from a spider's web, and along these routes ran buses. The town centre was therefore the only place which was accessible to all parts of its catchment area known as the point of maximum accessibility. As Figure 3.1 shows, the radial street pattern of most towns encourages movement into the town centre (radial flows) at the expense of sideways movements (circumferential or tangential flows). Using the analogy of the clock, someone living in a suburb at say five o'clock wishing to visit a friend at three o'clock could well find it quicker and more efficient to take a bus into the town centre and change rather than try to go direct.

Because of this structure, retailers and other service providers located in the town centre because it was the only place which made them accessible to the whole town and its catchment area. If that was their market, then they effectively had no choice but to compete for a site in the town centre, often paying high rents for the privilege. This is the model described in the text books, the basis for the retail hierarchy and the theory of central place. The only challenge to the town centre's monopoly was centres of different levels of the hierarchy. A distant larger centre might cream off a share of the more specialist comparison purchases, such as jewellery, and a suburban centre might intercept food and other more routine shopping.

Widespread car ownership completely transformed this picture. In fact it stood it on its head. Far from being the most accessible location, the

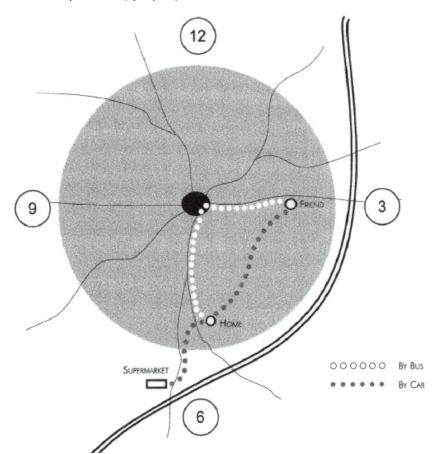

Figure 3.1 The urban clock

town centre became the least accessible. Far from shoppers choosing the town centre because of its accessibility, they continue to use it despite its inaccessibility because that was where the concentration of shops was found. The spider's web road pattern concentrates congestion in the town centre because such a high proportion of journeys within the town are forced to pass through it. To use the word 'revolution' is literally correct and no exaggeration: the wheel spun 180°.

The reason is the size of the car and its appetite for space. Roads and car parks make huge demands for space. They can occupy a quarter of the total space and sometimes more in areas of high density where land is dear. This can be seen best by comparing the space taken by a car, which often contains only one person, with the space taken by a public transport user. For a start the pedestrian needs no car park. They might leave a coat in a cloakroom, but apart from that the vehicles they use do not require storage in the town centre. Think of the space occupied by six cars in a car park. When turning and circulation areas are added, a parked car occupies about 300 to 350 ft^2 per car. Six cars therefore need about 2,000 ft^2 (180 m^2).

Now think of a lift with a capacity of six persons. It is unlikely to be bigger than 12 ft^2, say 3 ft by 4 ft. People are often packed in tighter than that in office lifts. Once in a car, therefore, our six require storage space at 2,000 ft^2, 167 times more than the 12 ft^2 which is all they need to be moved, say, from the multi-storey car park to the level of the shops in the adjacent centre. Of course the comparison is a crude one because car parking spaces are used by several different cars during the day, but it gives some idea of the order of magnitude of the different demands of space made by car users. A similar comparison could be made of the space occupied by the travelling car, complete with the space separating moving vehicles from each other, and the bus which occupies little more space but might be carrying 30 or 40 people.

From the shoppers' point of view, what they see is congestion and parking charges. Even after the heroic efforts made to accommodate the car with multi-storey car parks and inner ring roads in the 1960s described in the previous chapter, the car shopper is still likely to find a trip to the town centre slow and congested and parking difficult and expensive. In other words the trip is made because the town centre still contains some shops and services which are not available out of town. But the trip is still made despite not because of accessibility and location. Even for non car-using shoppers, the pressure from cars decreases accessibility. The bus passenger is delayed because of the congestion caused by cars, and the pedestrian often finds it necessary to negotiate windswept bridges or foul-smelling underpasses to reach the shops.

Obviously the problems that town centres have with cars has encouraged pressure for out-of-town development. To return to the analogy of the clock, the person living in the suburbs at five o'clock gains greatly by using a car to take them to three o'clock. Similarly any movement outwards to edge-of-town retail development now becomes possible, taking often less time than the former trip to the town centre. In fact as inner city populations have declined the average distance from house to town centre has risen in most catchment areas. This has further increased the accessibility of out-of-town development compared to the town centre.

The impact of widespread car ownership has been with us for 20 years, and though traffic is set to double in the next two decades or so, to some extent the novelty has warn off. There is a ceiling to car usage. In America there are now more cars taxed and insured than there are driving licences. One result of this is a decrease in the relative advantages of out-of-town. Values there now often match those in town centres and congestion is becoming a problem for retail parks and superstores as well as town centres. It is a phenomenon known in America as suburban gridlock, and as Dunphy

(1997) showed in his study of American transportation, it is caused by a failure to build roads on the expanding outer edges of cities on a scale to match the growth in housing and employment.

Of course low density living and car dependency has major drawbacks. Those without cars (the old, the young and the disabled) are at a severe disadvantage, and those with cars can spend much time chauffeuring those without. It is also widely felt that a sense of community is difficult to maintain in car dependent societies. It should be remembered too that it is possible to resist the corrosive effect of the car on the town centre by strong and consistent planning laws and by good public transport. Germany is probably the best model here. The Germans have a higher level of car ownership than Britain, but they seem happy to leave their cars at home to shop in the city centre because out-of-town development has been severely limited and they can ride a network of cheap efficient trams. Despite this it should be accepted that the car offers great advantages to the shopper, not least in carrying goods home. Although the German example shows that its effect can be contained it is nevertheless a constant pressure pushing towards decentralisation and low density land use.

The revolution in distribution

The effect of the car on location has been revolutionary even though it was caused by one single invention. The revolution in distribution, on the other hand, though equally dramatic is the result of a series of changes in technology ranging from the simple wooden pallet to the latest in computers. The dramatic changes in ports and docks which led to the whole of London docks being replaced by a container port at Tilbury resulted from the adoption of standard sizes for the steel container, and, of at least equal importance, the wooden pallet. Goods were packed on pallets which fitted inside containers. The goods were now delivered by ship to container ports where they were loaded straight on to lorries, or arrived direct via roll-on-roll-off ferries. The result was the elimination of goods handling and storage at the wharfside, and the job of the docker disappeared.

Quite suddenly the major ports, New York as much as London, became obsolete and were replaced with a few container ports and specialist terminals to handle bulk commodities (oil, timber, grain, ore, etc.). The space occupied by the new container ports and terminals was small compared with what they replaced and waterfront employment fell to only a fraction of what had been there before. All the new ports needed was good road access, deep water quays and a big area of hard standing on which containers could be sorted and loaded. The throughput is so much faster than before that the space needed is much less even though the volume is higher. The overall effect of the change has been to shorten and simplify the supply chain. Intermediaries such as wholesalers have been bypassed as big retailers have dealt directly with manufacturers. The very concept of storage itself has been eliminated. The Institute of Logistics estimated that in 1983 distribution amounted to 12.3 per cent of retailers' costs, but by 1992 this had dropped to only 4.7 per cent.

In locational terms the effect has been one of concentration. The trend to fewer but bigger which runs like a thread through this book, applies here with a vengeance. The best illustration of the concentration of modern warehousing comes from the supermarket sector. In 1995, for example, Tesco opened a 525,000 ft^2 (50,000 m^2) warehouse on the Welsh side of the Severn Bridge. It supplies 90 supermarkets in Wales and the West of England and has a capacity for 42,000 pallets. A lorry can carry 25 pallets. The four big operators (J. Sainsbury, Tesco, Asda and Safeway) between them supply around half the nation's food, and to do this they require a total of only 59 warehouses operating out of 14.6 million ft^2 (1.35 million m^2). To put this in context the Valuation Office Agency recorded 82,683 warehouses in England and Wales in 1994 with a total floor area of 886 million ft^2 (82 million m^2) giving an average (mean) size of 10,700 ft^2 (1,000 m^2). The 59 supermarket warehouses, therefore, with an average size of 247,000 ft^2 (23,000 m^2) account for only 1.6 per cent of the total warehouse space, yet from this they can supply half the country's food.

Much of the concentration in warehousing, described in chapter 10, has resulted from the arrival of 'high bay' warehouses in the last 15 years of the century. These giants have floor to ceiling heights of over 30 feet (10 m), equivalent to a three-storey office block and accommodate seven or more tiers of racking each containing room for a pallet. Advances in forklift truck design to reach these great heights has made this possible, together with floors of billiard table flatness designed to withstand great weight. The distribution revolution has resulted in most goods passing from factory or port of entry to shop with only one intermediate stage where they are sorted rather than stored. Partly that this has been made possible by the physical changes mentioned (pallet, container, forklift truck and high bay warehouse) but of equal importance has been the role of the computer.

Change has centred on bar-coding and EPOS (Electronic Point Of Sale). With bar-coding, each retail 'line' has a separate machine-readable code and as a superstore carries tens of thousands of lines, this has enormous potential. When an item is sold the EPOS system at the checkout passes the information back up the supply chain. The supermarket knows instantly what needs replacing on the shelf and ordering from the warehouse and eventually from the manufacturer too. Bar-coding and EPOS are marvels at improving efficiency and reducing waste. Suppose a manufacturer makes blue socks and red socks, and suppose the red socks are selling better than the blue, then the sooner this is known the sooner the division of the production between the two colours can be altered. In this example the danger of making unwanted blue socks while failing to make enough red socks can be avoided because bar-coding and EPOS provide rapid market intelligence.

The changes in distribution technology have been truly revolutionary but they all point in the same direction. They automate the flow of goods, they reduce the number of stages between manufacture and sale and they lead to a concentration into fewer larger units giving a huge increase in productivity. In terms of land use and location there have been two main effects. The first has been to free up large areas of inner city land which had previously been used for railway marshalling yards, docks and warehouses. The second effect has been to allow the distribution industry itself to follow a new and efficient policy of location. There are a number of sophisticated models which have been developed using operational research techniques to provide optimum locational strategies for different types of distribution.

Overall the result has been to lead to a concentration of high bay warehouses in the south Midlands, and this is described in chapter 10. Land values in the south Midlands are less than half the level around the M25. A distribution industry which has been freed to locate where it chooses can introduce the cost of land into its optimising models. It is time now to look at land values and how they vary, the subject of the next chapter.

4 The importance of value

A world of extremes

The price of property (its value) and the cost of occupying it (its rent) vary enormously depending on its location. Taking roughly similar buildings the variation resulting from location can often be as much as 10:1 and on occasion can rise as high 50:1 or 70:1. This means that certain shops and office users find it pays them to bid many times more for a desirable rather than an undesirable location. Shops in particular are ultra-sensitive to location. In introducing the role of value in location, the best place to start is to ask why this should be so. Businesses do not pay 10 times more for a site than they need to unless they believe the benefits justify it.

A shop in Oxford Street, say, near Oxford Circus among the most expensive areas, might pay around £450 or more in rent per year for each square foot of frontage it occupied. This is a huge amount. It means that a shop of 2,000 ft², a normal sized shop, though fairly modest for Oxford Street, would have to pay a total of just over £500,000 per year for the privilege of occupying a small plot of land. The figure is less than 450 times 2,000, i.e. £900,000 because the back part of the shop is charged at a lower rent using a standard formula. The formula involves halving the rent per square foot in each zone as distance from the frontage increases. The front zone is known as zone A, and rents quoted are for zone A unless they are for big stores where a figure for square foot rent overall is used. Zone A has a depth of 30 ft in central London and Scotland, but 20 ft elsewhere. The results of the formula in the example of a shop in Oxford Street are shown in Table 4.1 and Figure 4.1.

In the case of the retailer the site is all he gets. Convention in Britain is that retailers pay their own shop-fitting costs, and they, rather than the landlord, pay for repairs and insurance, not to mention overheads like lighting and heating. Most shops have total sales which are less than the rent which our Oxford Street shop pays. Put another way, if rent took 10 per cent of sales then the shop would need sales of £5 million from its small premises to survive. Rent as a proportion of sales varies widely from 2–3 per cent for supermarkets up to as high as 15 per cent. A common rule of thumb is

Table 4.1 Formula for calculating shop rent*

	Frontage (ft)	Depth (ft)	Area (ft²)	Rent (per ft²)	Total rent
Zone A	25	30	750	£450	£337,500
Zone B	25	30	750	£225	£168,750
Zone C	25	20	500	£112.50	£56,250
Total	**25**	**80**	**2,000**		**£562,500**

Note
*Assuming a shop in Oxford Street with 25 ft frontage and a total area of 2,000 ft².

7 per cent for clothing shops, and 10 per cent is fairly common. Retailing is the most sensitive of all property uses to differences in location. No other activity pays even a quarter the level of top shop rents. The nearest rivals are City of London offices and luxury West End flats where equivalent rents in the £50–£80 range are found.

Retailing is also the most sensitive of any land use to minor variations in location. If the shopkeeper in Oxford Street felt like throwing a stone down the side street beside his shop, he would find rents plummeting as rapidly as the stone. Within 30 yds they might fall from £450 to less than £150, or more than £10 for every pace of the few shoppers who bother to turn down the side street. Similarly in the City there are very steep rental gradients to the north and east of the highest rented area. In particular the prime Broadgate development round Liverpool Street Station sits near to offices to the north along Bishopsgate of much lower rents. Rents drop from around £60 to £20 per square foot in no more than 200 yds.

Let us look now at why some retailers and office users pay above average rents to get the premises they want and to ask, as retailers often do of their own customers, why pay more? The reason is they are buying accessibility to their customers which in their view is worth the price. Shoppers do not want to walk more than they have to, and are seeking to search for the goods they want with a minimum of effort. There is an advantage to them if shops cluster together and they reward the retailer who is easy to reach by spending money there. There is an element of circularity here as shoppers go where shops are clustered and shops cluster because shoppers are concentrated there. Despite this, certain types of retailer have learnt that their sales vary enormously according to how easy they make it for their customers to reach them. This argument does not apply to all types of retailers. Some, like supermarkets, are destinations in themselves and need only be reasonably accessible and prominently located to prosper. Others however, particularly those selling goods like clothing where the shopper wants to make a comparison before buying, need high levels of visibility and accessibility. The shopper is prepared to spend only a limited time on searching and comparison and it is vital to these retailers that their shops

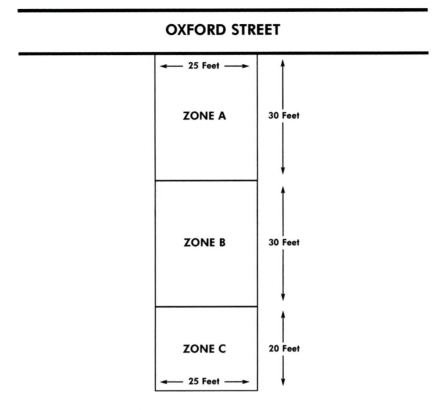

Figure 4.1 Diagram of shop zones

should be included in that limited search process. They can achieve this by locating in the prime retail position and are prepared to pay a high price to gain a position there.

Obviously high rent can only be paid for by a combination of high sales and high margins (or mark-ups) on goods sold. Multiples that tend to operate a standard level of mark-up are reluctant to pay top Oxford Street rents, and in fact Oxford Street has a lower proportion of multiples than the centres of most British towns. The less well-known shops that thrive at this extreme top of the rental range are a rarefied breed and one is reminded of some exotic creature which survives by exploiting a niche in the environment. Most sell gifts or cheaper items of clothing such as t-shirts, belts and fashion accessories, and there were in 1997 no less than 155 shops in Oxford Street selling clothing and footwear. Their market appears to be mainly foreigners and visitors from other parts of Britain who are looking for gifts to take home. Their mark-up is very high and they survive by being at the front end of fashion. Not surprisingly they frequently change hands and between 1990 and 1997 47 per cent of all Oxford Street shops changed their fascias (the name over the door).

The reason why businesses pay high office rent is different. Prime office space in the City or the high rented area in the West End is particularly attractive to certain types of business and this is discussed fully in the chapter on office location (chapter 9). Here we need only point out that the top rent payers are not, as some might assume, the offices of the big corporations, but rather the lawyers, accountants and advertising agents who service them. Overseas firms opening an office in London are usually prepared to pay top rents so that their presence has high visibility. What these two groups have in common is the desire to be accessible to their customers and to keep in touch with movements in their market.

How valuation works

There are four measures of value. First, and most important are rents. The examples of extremes in the previous section are all given in terms of rents. Rent is the income which a property produces and is the engine that drives the financial property train. The other three measures, value itself, rate of return and yield are no more than carriages on that train. Rent is particularly valuable for analysis because it enables comparisons to be made on a like-for-like basis between the various types of commercial property like shops and offices in different locations. The value of a property refers to the capital value of the freehold, but as properties vary in terms of their position, age, size, quality of construction and many other features they are unique and their values harder to use for comparison.

The third measure, yield, is the ratio of rent to value. Thus a property with a value of £200,000 producing a rental value of £20,000 p.a. would have a yield of 10 per cent. Yield shows the value placed by the investment market on the stream of rental income. The lower the yield the more highly the income is valued, implying that it is secure, liquid and expected to grow. It is the property equivalent of the price earnings ratio for shares. The fourth measure is the rate of return. This allows changes in value arising from changes in yield to be added to the income produced by the property. It produces an income-plus-capital return on the investor's holding. Rental values and yield are set by the open market, value and rate of return are calculated from rents and yields.

Rent is paid where the occupier of the property is not also the owner of the freehold. In Britain the majority of houses and farms are owner-occupied and so no rent changes hand. The majority of better quality commercial property on the other hand – shops, offices and warehouses – have different landlords and tenants. To a greater extent than applies in America or continental Europe, Britain has a group of professional companies and financial institutions who are effectively professional landlords, treating property as a form of investment like any other and concentrating their holdings on better quality, modern and well-located property. Even where big multiple retailers like Boots the Chemist own the freehold of many of their own branches, it is common for them to charge themselves the open market rent as a way to gauge the profitability of each branch. The value of a property is what it would fetch in the open market if offered by a willing seller and properly marketed. In this respect property is the same as any other asset. It is different, however, in that there is often the need to know the value of a property without actually selling it, and the evidence of actual sales which can be used as comparables are few and varied. That is why there is the need for the profession of valuer.

It is worth mentioning two other characteristics of property which set it apart. Tenants sign a lease with a fixed length, now commonly 15–25 years, longer for older leases. The rent they pay is reviewed every few years, commonly five years, and adjusted to the open market level. In a rising, inflationary market this means that the rent the tenant pays remains flat for five years but then goes up with a bump. This can be enough to cause shops to close, and the occupancy of streets can change quite suddenly if rent reviews occur at the same time, particularly if this happens in a recession. The second characteristic is multiple ownership. Not only do both landlord and tenant have an interest in their property's value, but so also does the bank which might have given a loan secured on that value. In many cases, there are also intermediate layers of ownership between landlord and tenant. Thus the landlord may have granted a 99-year lease to A who in turn gave a 50-year lease to B who let it to the occupier. All these interests have a value.

Valuers work in terms of rents and yields, and the property market measures itself in terms of them. Surveyors who are active in the market know the current level of prime shop rents in, say, Swindon or offices in Slough and whether these are rising, static or falling. The valuer takes these market measures and then adjusts them to the circumstances of the specific property in order to calculate a value. Value is therefore derived from rents and yields.

People complain that valuation is subjective and that valuers often disagree. Indeed valuers can spend much of their time arguing in front of arbitrators and Rent Tribunals. They contrast this with the stock market where precise share prices are published each day. But this problem is not as bad as it first appears. Partly this is because much of the disagreement over value happens in the adjustment process. There is less argument over the prime rents and yields which form the valuer's starting point. Also the apparent precision of share prices is largely spurious, particularly among smaller quoted companies. Whatever it may say in the newspaper, a major buyer or seller in a small company is unlikely to find that they can deal at the quoted price. Their interest will itself make the price change. The patterns of rents described in this chapter are based on market valuations, but they are based on a consistent data set and give as accurate a measure as any in the imprecise world of asset valuation.

Residual value

Rent has played a role in classic economic theory for the past two centuries. Ricardo who is the founding father of rent theory, wrote his leading work in 1817, but although the industrial revolution was well underway by then, he and J. S. Mill and Marx who followed him saw rent in terms of agriculture. Classic theory sees rent as one of the great abstract nouns of economics, along with utility, labour, marginal output and many others. It does, however, produce two ideas relevant to location. One of these is the idea that values fall with distance from the city (considered in chapter 9) and the other concerns the concept of residual value.

Rent is considered to be a residual because its level is what an occupier is prepared to bid to have the right to use the property, that is the site and the building on it, to make money. If we remove the effect of the building, what remains and the way it varies between sites is a pure expression of location. Indeed rent is the best measure of location there is. It is a sensitive, quantifiable measure of the demand for a particular location, and, if properly measured, is more focused and more accurate than other indicators of demand such as actual occupancy and planning applications.

Theoretically, the level of rent should be whatever the occupier can afford to pay after allowing himself a reasonable profit. If something happens to increase the profitability of the location then the extra profit goes to the landlord in the form of extra rent and not to the occupier, even if the occupier has himself caused the increase in profit. This is the sense in which rent is a residual. If, for example, a site becomes more valuable because Marks & Spencer decide to open next door, then it is the landlord who receives the benefit as a windfall. This seems unfair because the creator of the extra value, Marks & Spencer, receives nothing. It is easy to understand, however, that retailers would compete by offering the landlord higher rent for the benefit of being sited next to Marks & Spencer.

The same thing happens if an exploration company finds oil on private land, and even in cases where the tenants themselves bring about increases in value through their own efforts, such as drainage undertaken by a tenant farmer. The landlord receives the benefit without needing to lift a finger. It was the idea that this residual value would naturally accrue to the landlord which was one of the driving forces of Marxism and it has stuck in the craw of left-wing parties ever since. It led Labour to tax development land and betterment in both the Attlee and Wilson governments. Betterment is the increase in private land values resulting from public expenditure, such as road building.

In terms of location, residual value influences both the density of development and the frequency of redevelopment. Residual values are highest in the most desirable locations where rents are highest such as City offices and Oxford Street shops. In order to increase the return on investment it pays the landlord to increase the rent received by raising

Table 4.2 Example of how high residual value encourages investment

	City office block		Northern warehouse	
	Before investment	*After*[1] *investment*	*Before investment*	*After*[1] *investment*
Building	£25m	£100m	£7.5m	£30m
Land[2]	£75m	£100m	£2.5m	£5m
Total value	£100m	£200m	£10.0m	£35m
Rental income[3]	£2.5m	£10m	£750,000	£3m
Return on value[4]	2.5%	5.0%	7.5%	8.6%

Notes
1 Freeholder replaces building with one four times the value.
2 The residual value of the land increases following the new investment in the building.
3 Assumes rental income is 10 per cent of building value in all cases.
4 Return on value is rental income as a percentage of value. In the case of the City office block this doubles following investment but for the northern warehouse it rises far less.

the physical density of development on the site, for example by building a skyscraper, and by raising the rent that can be charged per square foot by redeveloping and modernising the property. Table 4.2 shows a theoretical example of this by comparing a City office with a north of England warehouse. In the former the land (the residual value) accounts for 75 per cent of the total value with the remaining 25 per cent taken by the building. With the warehouse, however, the building accounts for 75 per cent of value and the land only 25 per cent. Now in each case the freeholder decides to invest by replacing the existing building by a big new building with a value four times the existing building. The table shows that the City office investment results in the return doubling from 2.5 per cent to 5 per cent. The warehouse investor in contrast receives only a meagre increase in return from 7.5 per cent to 8.6 per cent, scarcely justifying the risk and the effort.

This crude and simplified example goes some way to explain why skyscrapers when they are grouped are found in high value city centres, and why in times of property boom such as 1987–90, tower cranes were seen in abundance in central London but were lacking in the towns and cities beyond the M25. The conclusion is that desirable locations with high rents attract frequent redevelopment and high density. In fact high density development commonly leads to greater unit cost of construction which makes the situation more complex than in the table. This is a problem which effects the skyscraper builders in America rather than the city centres of Europe where height restrictions are widespread. High density development in Europe is very much a relative term.

Patterns of rent

A number of chartered surveying firms hold databases of rents and yields covering the various types of commercial property across the country. That

Table 4.3 Shop rental growth by region: May 1977–November 1999

Above average		Below average	
Wales	992	South East	676
East Anglia	883	South West	664
West Midlands	864	Scotland	646
North West	713	Yorks and Humberside	604
London	705	East Midlands	549
North	699		
London and the South East			
Central London	827	South East	676
		Outer Suburbs	617
		Inner Suburbs	535

Source: Hillier Parker (1994)

Notes
Index: May 1977 = 100
GB average = 698

of CB Hillier Parker which is the main source used here was set up in 1977 and today covers 1,293 rent points where rental values and yields are recorded annually or more frequently. The data refer mainly to prime rents, that is open market rental value of a property of good specification. Table 4.3 shows the regional pattern of shop rental growth over a 22-year period from 1977 to 1999. There is no obvious pattern, with Wales performing best and East Midlands least well. There is certainly no evidence of the south outdoing the north, with the South-East region excluding Greater London doing no better than the average. Perhaps the most significant finding is the inner city effect, with the inner suburbs performing worst and rental growth then rising with distance from central London. It is notable that even the outer London suburbs and the rest of the South-East failed to achieve the growth of the national average.

Table 4.4 shows equivalent figures for offices and industrial property. As with shops, offices and industrial performed poorly in the London suburbs, under performing both the South-East and Central London. There would appear to be evidence here of the American idea of the doughnut effect of inner city decline between a thriving city centre and an expanding region surrounding the city. What many people find remarkable is that an area as vast as the London suburbs, lying at the heart of the most prosperous quarter of the country should have done so badly in terms of offices and shops, though not in the case of industrials. Alone of the three property types considered here, industrials show a clear north–south pattern with the long-term pattern of rental growth falling with distance from London.

The figures for offices also show low growth for Central City, the area which in 1977 was the highest rented in the country. In fact it was during this period that the West End replaced the City as the highest rented area (in the late 1980s). The failure of the Central City even to match the national

Table 4.4 Offices and industrial rental growth by region: May 1997–November 1999

	Above average			Below average	
Offices	West Midlands	838	East Midlands	461
	South West	702	South East	430
	North West	543	Yorks and Humberside	392
	London	516	East Anglia	377
	Scotland	514	North	294
	Wales	479		
London	West End	782	Central City	417
	Holborn etc.	634	Suburban	361
	Fringe City	502		
Industrial	South East	462	West Midlands	389
	London	443	Wales	364
	South West	438	Scotland	358
	East Midlands	437	North West	343
	East Anglia	424	North	300
			Yorks and Humberside	286

Source: Hillier Parker (1994)

Notes
Index: May 1977 = 100
GB average for offices = 463
GB average for industrial = 390

average rate of growth stands in stark contrast to the other parts of central London, all of which performed well, including Fringe City. It is worth noting here that the core of the City is hampered from major redevelopment by a combination of planning controls, listed buildings and fragmented land ownership often in the hands of non-commercial bodies like City livery companies. The 22 years covered by the table include the development of Canary Wharf in Docklands and the redevelopment of Liverpool Street Station with the appearance of Broadgate nearby. These occurred because of the failure of the prime City area to keep pace with the demand for large floor plate modern offices.

Rent mountains

One of the purposes of this chapter on values is to give some indication of the usefulness of rental value data in location studies. Academics in particular have neglected rents as a source, possibly because they feel they lack the rigour required from data, and possibly because they have not been readily available. Rental values are only estimates, but if we accept a margin of error they can reveal broad patterns and underlying pressures for change in a way no other source can. Let us start with the hypothesis from classical economics that rents fall with distance from the city centre, mentioned earlier. In broad terms it is mostly true, but there are exceptions, and

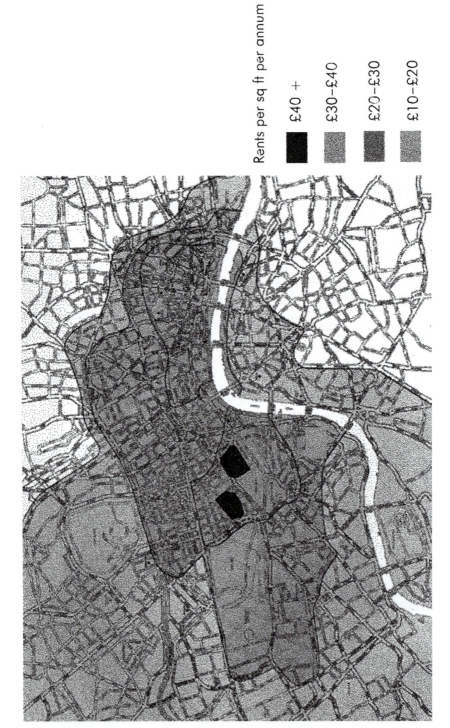

Rents per sq ft per annum

■ £40 +

▨ £30–£40

▨ £20–£30

▨ £10–£20

Figure 4.2 Central London office rent contour map

the changes which are described throughout this book suggest that the exceptions are becoming more widespread as the pattern becomes more complex. The doughnut phenomenon of inner city decay, for example, can lead to an irregular fall in values with distance as was demonstrated in the previous section. The deep-seated move to decentralisation also serves to complicate the pattern as the difference between centre and periphery decreases. The possession of up-to-date rental value data means that these trends can be revealed in a way that would not otherwise be possible.

In most situations, however, the point of highest rental value lies in the city centre on top of a mountain of rental value. In the case of shops there is a prime position known as the 100 per cent pitch which might be considered as a small plateau at the peak of the mountain. This area varies according to the street pattern of individual towns, but is usually no more than 20 to 100 metres long. Surrounding this, rental values fall dramatically to the edge of the town centre and then more slowly outwards towards the edge of the town. In a good secondary position within the town centre, good enough to attract non-fashion national retailers such as Halfords, rents might be a third of the prime level and in minor streets they could be as low as 10–20 per cent.

The shape of the rental mountain varies between cities for historic reasons. Some, like Bournemouth, Sheffield, Brighton, and Grimsby are weakened by having split centres which shoppers and retailers find confusing. In fact it is noticeable in the four examples quoted the stronger half of the centre has strengthened at the expense of the weaker over the last 30 years. In cities where the centre was rebuilt after bombing such as Plymouth, Hull and Bristol, together with the post-war new towns, the high rent plateau tends to be larger and hence top rents lower than in cities such as Leicester where the medieval street pattern continues to produce a tight 100 per cent pitch. Generally, however, a city centre seems to function most effectively where there is a clear 100 per cent pitch which is big enough to give the multiples the space they need, and where there are attractive fringe zones for specialist shops. It has been the failure of the prime areas to provide adequate space for the multiples which has led to the development of comprehensive shopping centres which have been grafted onto the centres of virtually all towns in the country of more than 50,000 population. Often the 100 per cent pitch then moves inside the centre, but, as developers have learned, shopping centres themselves can have secondary areas. The location of the 100 per cent pitch is surprisingly durable and there are limits to what development can do to alter the shape of the mountain.

For offices the pattern is similar, although offices are less locationally sensitive than shops. Unlike shops, however, it is possible to talk of a national rent mountain centred on London and the area to the west (see chapter 9). The contour map of Central London office rents (Figure 4.2) shows how far the City contrasts with its low rent hinterland to the north, south and east, in fact in every direction except west. To the west the high

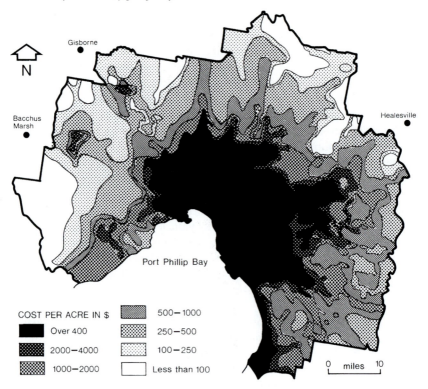

Figure 4.3 Melbourne region: land values 1969–1970
          (Original produced by the Metropolitan Board of Works, Melbourne, 1971.)
          Source: Howes (1980)

rented area fans out to cover a wide part of the West End and thence on to West London. With industrial rents there is also a national rent mountain as with offices. Analysis of the rent contour map (see chapter 10) suggests that distance from London has a stronger explanatory power over rents than other rather obvious factors such as proximity to a port or a motorway.

In 1980 Howes published his book, *Value Maps* showing different techniques for mapping land values. Most of these showed rent or value mountains as described here. As illustrations we reproduce an example showing contour lines of land prices in Melbourne (Figure 4.3).

The move west

One of the most significant trends revealed by rental analysis has been the move of high values westward through Central London. It is possible to argue that the centre of gravity of London has moved westward from the

Romans' London Bridge and William the Conqueror's Tower to the centre of government in Westminster in medieval times and beyond. In rental value terms, however, there has been a remarkable change in the last 20 years. Since 1980 there has been a distinct shift to the west of the highest rented area for both shops and offices. In retail the area of top rents has changed from Oxford Circus to Knightsbridge. This occurred during the 1980s and involved a big geographical move across Hyde Park. This is supported by other evidence such as the earlier closure of Gamages to the east in Holborn which rivalled Selfridges in its day, and by the growth of designer fashion in Sloane Street (off Knightsbridge) and the increasing importance of Kings Road, Chelsea and High Street, Kensington, both in the western extreme of Central London.

With offices a similar shift has occurred with Mayfair and St James's in the West End overtaking the Bank of England as the highest rented area in the country. According to the Hillier Parker rent index this change took place in the late 1980s. The significance of the development of the West End as an office centre is discussed in chapter 9, but it is still remarkable when one remembers that the great weight of the office stock as well as centuries of tradition is concentrated in the City. In some respects there is a similarity between shops and offices because Knightsbridge is tiny in terms of floor space when compared with Oxford Street. Apart from Harrods, Knightsbridge barely qualifies as a major retail street. It has no branch of Marks & Spencer for instance. In both shops and offices, therefore, the area of greatest physical concentration is no longer the area of highest value, and for both the move west occurred at roughly the same time.

There is an interesting parallel in this with New York. Wall Street or Downtown at the southern tip of Manhattan ceased to be the area of highest rents when it was overtaken by Midtown in the 40/50th Streets, an area equivalent to the West End. In shopping too there has been a steady move northwards from Macys on 34th Street to 42nd Street and 57th Street on 5th Avenue, with retail growth in the Upper Eastside (68th to 70th Street and Madison), the high income residential area, playing the role of Sloane Street. So we have seen top value shops and offices in the two great cities of London and New York both moving quite markedly away from the traditional area towards the area of high income housing. This is another trend which forms a theme in the location evidence presented in this book. High income housing begins by being separated from the main commercial area, but over time that commercial area moves towards it.

Areas of growth

As the reader might expect, there is a rough correlation between a town's prime rent and its population. Top rents in Central London are higher than those in Birmingham (by a ratio of around 2 : 1) which in turn are higher than Worcester (again roughly 2 : 1), and so down to small market towns.

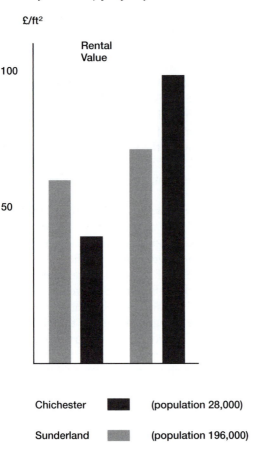

£/ft²

Figure 4.4 Rental values in Chichester and Sunderland
Crown copyright is reproduced with the permission of the Controller of
Her Majesty's Stationery Office.
Source: CB Hillier Parker

It is probably more interesting, however, to see where population gives a misleading guide to shop rents, because there are signs that the rent/population correlation is getting weaker. In the mid-1980s, the top five cities outside London in terms of shop rents were found among the great regional cities (Edinburgh, Glasgow, Newcastle, Leeds, Sheffield, Manchester, Liverpool and Birmingham). A similar list in the late 1990s would include out-of-town centres such as Brent Cross (north London), Meadowhall (Sheffield), Lakeside (Thurrock) and the Metro Centre (Newcastle). It would also show that attractive tourist towns like York, Chester, Bath, Oxford and Cambridge could demand rents of a level similar to those of big cities several times their size.

Recent years have seen the emergence of smaller towns in the prosperous ring around London, such as Guildford and Horsham with rents and values surpassing those of larger northern towns. Figure 4.4 shows how the relationship between Chichester and Sunderland has changed since 1984. Sunderland with a population of 196,000 dwarfs Chichester's 28,000 (a ratio of 7 : 1), and up until the mid-1980s rents and values there were higher, though by less than 2 : 1. By 1993 the scene had changed and Chichester's rents had clearly overtaken those of Sunderland.

Central London provides another interesting example of rental growth. As well as the emergence of Knightsbridge there has been even more rapid growth in Covent Garden. When the fruit and vegetable market moved out, Covent Garden was converted to a speciality or festival shopping centre. For over two decades it was a modest success, but it has only been during the 1990s that rents have soared. They grew by 150 per cent between 1990 and 1997 to a level matching parts of Oxford Street.

With the arrival of Marks & Spencer in Long Acre and many fashion multiples taking space in the surrounding streets, Covent Garden seems to be changing into a more conventional shopping area. It is a type of metamorphosis hitherto unmatched in British retail history. One theory is that if colourful independent retailers make a success of an unconventional retail location they attract visitors. These then attract the multiples that disappoint the visitors who then go looking for Camden Lock or the next place to which the independents have been forced to migrate by high rents. If this theory is correct then the multiples are doomed to spend their time catching up like the back half of a caterpillar forever following the front.

Pigs and pythons

The ripple hypothesis states that change begins in Central London and its effect then ripples outward to the rest of the country in the same way as the ripples created by a stone dropped in a pond. It is a reasonable idea since decision-makers in commercial property are concentrated in Central London, and it is supported by the work of Mean on the housing market described in chapter 11 later. He found that ripples began when interest rates changed with a greater initial reaction in London because mortgages were proportionately higher. There would appear to be non property evidence in support of the hypothesis for both population and politics. The area of maximum population growth lies in the ring round London which is slowly rippling outwards and currently stands at a radius of around 100 miles. At the moment it runs from Suffolk through the South Midlands to Somerset and Dorset. In the 1930s this ring was in outer London and in the 1950s and the 1960s in the Home Counties.

In voting patterns too, it is possible to see a similar if more complex picture. The way people vote forms a good proxy for social mix and has well reflected the growing polarisation between rural affluence (Tory) and big

Table 4.5 Shop value growth in boom and slump by type of town

| | Shop rental growth greater (+) or less (−) than the national average | |
	1984–88	1988–92
1 London boroughs	+	−
2 Metropolitan districts	−	+
3 Non-metropolitan districts	−	+
4 Industrial districts	−	+
5 New towns	+	−
6 Resort, port, retirement	+	−
7 Other, mixed	+	−
8 More remote, mainly rural	+	−

Data: OPCS Classification of Local Authority Districts; Hillier Parker data on shop rents in November of relevant year for 40 towns
Source: Urbed *et al.* (1994), page 161

city decay (Labour). Recent voting patterns support the idea of alternating decentralising rings of affluence. Thus a cross section starting in Central London would begin with Conservative dominance spreading out into adjoining traditional Labour areas such as Battersea, while Labour's vote has grown in outer suburbs like Norwood, Southgate, Harrow and Twickenham which used to be safe Conservative seats. The evidence from commercial property is not so clear. A major report on economic and property cycles commissioned by the RICS in 1994 found little sign of a ripple effect running outward from London through the regions. The findings of Henneberry, whose study is discussed in the final section of this chapter, suggests a contrast between London on the one hand and the rest of the country on the other, rather than a clear pattern of rents rippling outwards in a regular pattern. Certainly the 22-year changes in regional rents as shown in Tables 4.3 and 4.5 do not suggest the ripple concept. Apart from the inner city dip mentioned, there appears to be no regional pattern at all.

One possible reason for this is the difficulty of measurement. Even a period as long as 22 years can be affected by the timing of the beginning and end dates. A different time period might have produced different results. A detailed study of shop rental growth undertaken by the author as part of a report to the government on town centre vitality found alternating patterns of growth between different types of shop and different regions (Urbed *et al.* 1994). What seems to happen is that recovery affects the more modern and prosperous towns and newer forms of retailing first, then percolates to the average before reviving the more depressed last. It has been compared to a pig passing through a python. This means that if you select a time period when the pig has reached the tail of the python you can produce results which show above average rental growth in the most depressed areas.

Table 4.6 Commercial property boom and slump in the late 1980s

		Shops	*Offices*	*Industrial*
Trough	May 1984	113	99	91
Peak	Nov 1989	226	185	146
Trough	May 1994	175	93	107

Source: Hillier Parker (1994)

Note
Rent Index adjusted for inflation: May 1977 = 100

In the long continuous growth in retail sales volume which occurred in the 1980s, the peak growth year for rental values for different types of shop was as follows:

1983 Retail warehouse and supermarkets
1984 Market towns
1985 Secondary positions in town centres
1986 Big city centres
1987 Inner city locations

In the town centre vitality report mentioned above an analysis of rent points grouped into types of geographic areas showed an alternating or reciprocal pattern of rental growth. As Table 4.5 shows, no area had above average growth for both time periods studied (boom and slump) and none had above average growth for neither. These results conform to the pig in the python model and suggest that for retail at any rate it is the type of shop and the type of town which affects the pattern of rental growth rather than the region itself and its distance from London.

For offices, however, there is some support for the ripple hypothesis. Offices, more than shops or industrial are subject to extreme cyclical variation. As Table 4.6 shows, even after adjusting for inflation office rents virtually doubled then halved in a decade, a degree of oscillation unusual in economics which shows the sensitivity of rent as a residual. As the boom arrived, it was Central London office rents which rose first and then showed the first signs of downturn as the slump followed. Table 4.7 compares the turning points of the two major property booms of the past 30 years. Annual data only are available for the early 1970s and their quality is lower, but in both cases the table shows Central City offices turning down before the national all offices average. For both there was a time when office rents in the country at large continued to rise after they had peaked in the Central City area.

Henneberry took the analysis of the same time period further. Using the same Hillier Parker data he published a chart showing the turning point by region, reproduced here as Table 4.8. This does show elements of a ripple

Table 4.7 Cyclical turning points: central city and all offices

| | | Percentage change in office rents | |
		Central city	All offices
Early 1970s	1973	107.2	57.4
(Twelve months to	1974	−8.0	11.5
May of each year)	1975	−21.3	−2.8
Late 1980s	May 1989	2.9	11.9
(Six months to	Nov 1989	0.2	8.4
date shown)	May 1990	−1.4	5.3
	Nov 1990	−8.8	−3.6

Source: Hillier Parker (1994)

Table 4.8 Cyclical turning points in the regions: office rents 1989–1991

| | 1989 | | 1990 | | | | 1991 | | |
	Q3	Q4	Q1	Q2	Q3	Q4	Q1	Q2	Q3
London		□							
South East				□					
East Anglia				□					
South West				□					
East Midlands				□					
West Midlands					□				
Wales				□					
Yorks and Humberside							□		
North West							□		
North									□
Scotland								□	

Source: Henneberry (1996)

effect with a turning point lagging that of London as a whole as distance from London increases. Of course this is for only one time period and the evidence is less clear at other stages of the economic cycle, or indeed for earlier cycles. The ripple hypothesis, therefore, so often failing to show itself in the data, is nevertheless one which is well worth keeping in mind.

Yields and investment

As we saw earlier, yield is the ratio of rent to value. It is the main measure used by the property investment market to assess the quality of the property and the value placed on the stream of rental income which it produces. The lower the yield the more highly the income is valued. Investors when considering what property to buy, and hence what yield they are prepared to bid, need to make a judgement on liquidity, security and growth. Liquidity covers the ease with which the property can be bought and sold,

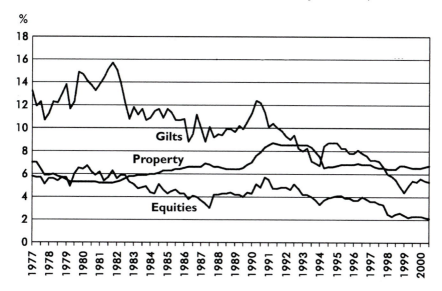

Figure 4.5 Investment yields
Source: CB Hillier Parker (1997)

its marketability. Security is concerned with the certainty that the property will be occupied and rent-producing, and growth deals with the prospects that the current rental value will grow.

Thus shop yields are lower than offices which are lower than industrials because, as Tables 4.3 and 4.4 indicate, rental growth in the past has been highest for shops and lowest for industrials. Vacancies (or voids) in prime positions are also felt to be highest for industrials and lowest for shops. Further, it is thought that the desire by retailers for a good town centre location will ensure a steady demand for property there to a greater degree than offices or industrials, although as we will see in the next chapters this easy assumption is perhaps not as safe as it once was.

The result of these perceptions by the investment market has been to produce the following levels of average yields for the points covered by the Hillier Parker Rent Index in November 1999:

Shops   6.5 per cent
Office   7.0 per cent
Industrials   7.7 per cent

This compares with a yield for gilts of 4.9 per cent and 2.3 per cent for equities, both much lower than investment property. In fact property yields have risen in the last 20 years relative to gilts and equities. For a time in the early 1980s property actually yielded less than equities, but since then equity

% change

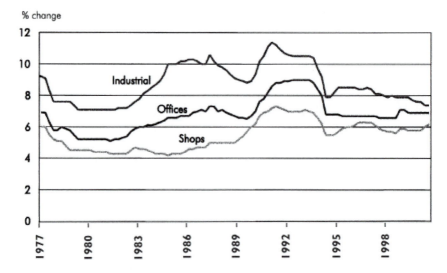

Figure 4.6 Property yields
          Source: CB Hillier Parker

dividends have grown faster than property rents while gilt yields have fallen
as the threat of inflation has receded. Bearing in mind that yields represent
the investment market's view of the quality of a stream of rental income and
therefore an implicit forecast, we can now look at how yields of similar types
of property vary by location. It is possible to produce contour maps for
yields as it is for rents, but strangely they produce an almost identical
picture to rent contour maps. The lowest yield and hence the highest value
is found where rents are highest. The correlation is close.

   One might naively expect that yields would be affected by growth rather
than the level of rent but this is not the case. High rent, after all, is an
indication of high growth sometime in the past but not necessarily growth
in the future. High rent is also taken to be an indication of demand and
hence gives reassurance to the investor who may be seeking liquidity and
security as well as growth. The result is that yield rises with distance from
London, and hence values fall. As for office and industrial rents there is an
area of high values and low yield to the west of London. The variation in
yields however varies less between places than it does for rents. Thus as we
saw above, Central London shops and office rents may be between three
and five times the level of rents in a substantial county town such as
Northampton or Derby, but in terms of yield the difference would be of the
order of 5 per cent against 8 per cent for offices and less for shops (5 per cent
against 6 per cent).

   One interesting change in yields has been in the attitude of investors to
Central London shops. Traditionally these were considered volatile, as

Table 4.9 Changing shop yields

	1972 May %	1985 May %	1999 Nov %
Central London	5.5	4.8	4.8
West Midlands	4.6	3.7	5.2
Average	5.4	4.8	6.5

Source: Hillier Parker (1994)

indeed in terms of rents they were. As a result Central London shop yields stood above the national average from 1972 to 1985 but since then they have fallen to well below the shops average (see Table 4.9). The dependence of Central London shops on visitors seems to be more acceptable than it once was, though rents there are still more volatile than in the country as a whole. As a general rule high rents are more volatile than lower rents, a fact which is accepted by the market as a price worth paying for desirable property, although it hardly fits with the security which investors may feel they are buying.

In the previous section, reference was made to Henneberry's work on the cyclical turning points of the different regions. He found that in the late 1980s boom yields turned a year before rents did, in 1989 rather than 1990. In other words investors were able to anticipate the downturn in the market for occupied property. He also found that unlike rents where the turning point rippled out from London, yields turned up at the same point in time across the whole country. This is probably because the investment market operates as a single national market controlled from London, whereas the rental market is affected by local conditions. He found that the late 1980s boom had seen little office development outside the South-East and that this was because the lag in rents rising was great enough for there to be insufficient time to trigger new development before the market turned down.

His study of the North-West region found that a development appraisal for a standard office block did not become profitable until the second half of 1989 just at the time when office yields were turning up across the country, although as we saw in Table 4.8, rents were still rising in the North-West. Not surprisingly in these circumstances no occupier would commit themselves and no bank would lend so that office development aborted before construction could start. The result was that unlike the early 1970s boom, the late 1980s boom was limited to new development within the Greater London area and the inner parts of the South-East.

5   A little theory

Forces in conflict

It is time now to look at the fundamental forces which drive location. As we saw in the second chapter they do not by themselves determine land use on the ground. That is the result of the interplay of politics and locational forces, with location often losing out to politics. In the third chapter we noted how radically the demand for space could alter following techno-logical change, and in the previous chapter we discussed value as a way of measuring locational demand. We are now ready to explore the fundamental forces which are at work beneath the surface. There are two of them and they are in conflict with each other.

First, there is the desire to concentrate, to cluster together, to gain the benefits of synergy and economies of scale and to give the customer the advantages of choice. This force leads to concentrations of high rent in small areas. The second force is the desire to disperse, to bring services closer to the customer, to decentralise and enjoy the benefits of more room and lower land costs. It is the exact opposite of the first force and its power has been strengthened enormously by the arrival of the car.

The two forces resemble the centrifugal and centripetal forces in astronomy and physics. The Earth spins around the Sun making it subject to a centrifugal force which on its own would send it out into deep space. At the same time the Earth is subject to a gravitational pull from the sun which on its own would cause the Earth to crash into the Sun. These two forces balance each other out at 93 million miles keeping the earth at a constant distance from the Sun. Every business is subject to both of these two forces and has to choose a strategy which they think is optimal for them. Put another way, they need to choose their own equivalent of the 93 million miles.

Most urban land users are subject with various amounts of sensitivity to these two forces. They do not apply where some quality of the land itself determines its use as in mining, and quarrying, and in agriculture their role is weak, but in most other cases it is possible to see the conflict at work. In other words people making location decisions have to take into account some factor other than the cost and availability of suitable sites, and that factor can be assessed in terms of the forces of concentration and dispersal.

Let us start with housing. Historically houses have tended to group into settlements, originally villages, then towns and now estates or neighbourhoods in larger urban areas. This is partly because man is a gregarious animal but also because it gives easier access to jobs and the increasingly wide range of services demanded. People buying houses, or house builders acting indirectly on their behalf, have a number of trade-offs to make. At the more strategic level they need to minimise the distance to the place of work and with the majority of women working this can now be quite complex, particularly if schools for the children also need to be considered. At the more local level there is a trade-off between space and convenience. This might typically involve a choice between a house without a garage within easy reach of the town centre and one on the edge of town with its own garage but beyond walking distance of services. This second house might offer more space as well as less noise and easier access to the countryside.

Long distance commuters pay a high price in terms of cost and time to enjoy the type of life they want and this often involves living within easy access of the countryside. They are choosing one end of the scale between space and convenience whereas a student sharing a tiny flat in Kensington has chosen the other. The concept of trade-off is explored more fully in chapter 11.

The second major land use category affected by our two conflicting forces is business use, which includes offices, warehouses and factories. At first sight the trend to decentralise appears to dominate, and it is certainly stronger than it once was. Changes in electronic communications and widespread car usage have certainly allowed business generally to behave in a more footloose way. The result has been to put a premium on attractive locations with research centres typically gravitating to the best. Despite this, the clustering force can still be seen at work. Some businesses locate in large centres regardless of the high rent and other costs because they want the profile and accessibility which they gain. They enjoy a similar advantage to clustering retailers in being near their competitors for the benefit of their customers. Thus law firms catering to the financial community need to be near the City. Business also benefits from being near pools of specialist labour to recruit their staff, and as a result rents are higher in the Thames Valley than in many parts of London.

The third and major example of the two forces at work is in retail. Retail is by far the most sensitive land use to even the smallest change in position as was shown in the previous chapter. As a result it is in retail location that most modelling and research has been carried out and this is explored below in the next three chapters. It is sufficient here to point out that the arguments that apply to retail apply also to other occupiers of similar space, particularly services. Strictly speaking retail refers to goods, and a shop selling services, such as a travel agent or a hairdresser is not classified as retailing. This can lead to trouble with different definitions of what is a shop being used by, for example, the Inland Revenue for rating purposes and the Department of Trade and Industry for retail sales. In locational terms, however, what matters is whether a business invites the public to come in off the street and spend money. Those that do include the growing catering and leisure industries as well as a whole range of services. So the workings of our two forces in the next part of this book which deals with retail, will also include these wider categories.

Central place theory

In the remainder of this book the two conflicting forces of clustering and dispersal are examined as they apply to various types of land use. Fundamental to this is the concept of hierarchy and this has a chapter to itself in chapter 6. But before starting on this it is worth looking briefly at past efforts to provide a theoretical underpinning of location. These mainly consist of attempts to 'explain' some observed phenomenon with arithmetic and geometric models. Thus the observation noted earlier that population density tends to fall with distance from the city centre has been modelled for over a century, notably by Clark who found that the decline fitted a negative exponential (i.e. gently curved) slope up to a distance of more than 300 miles. Peter Haggett in *Locational Analysis in Human Geography* reviews a whole range of models covering networks, distance decay, diffusion, the rank–size rule of city populations, gravity and central place. In this chapter we examine models covering central place theory, gravity and store location models.

Interestingly the pioneers are predominantly German. Thus there was von Thünen on agricultural location in the last century, Weber on industrial location in the early twentieth century, while the pioneers of central place theory, Christaller and Lösch are both German. The pattern among the academic community is then to debate the extent to which the theory of the model actually works in practice and to see if it can be modified to make it fit better. Models in the social sciences of which locational models form a part lack the precision of models on the physical sciences and are of value only if they offer a workable approximation to reality and help in the understanding of the force at work. Central place theory has had a profound effect on urban planning and government retail policy because its emphasis on the concept of the urban and retail hierarchy has so influenced the thinking of planners. As recently as 1996 a report by the Oxford Institute of Retail Management and Jones Lang Wootton reported that in Germany:

> land use plans have required retail development to be organised in a hierarchy of central location:
> - demand for essential supplies should be met in neighbourhood centres

- demand for non-essential specialised goods should be met in medium sized centres
- demand for fashion and luxury goods should be met in city centres.

In other words exactly following the hierarchy demanded by central place theory. It is one of the central themes of this book that widespread car use for shopping has made central place theory obsolete, and indeed it plays no part in the work of consultants advising retailers and shopping centre developers of location. This situation has been in existence for at least two decades and it is an indication of its power that it continues to dominate policy long after there has ceased to be empirical support for it. Let us briefly examine what the theory consists of. Stephen Brown has written a number of papers on the subject and in 1993 published what amounts to an historiography of what he described as the 'extraordinarily voluminous' literature on central place theory. He summarises the theory as follows:

> The traditional central place model predicts that, due to the increasing cost of transport, demand for any good declines regularly with distance from the source of supply. Beyond a certain point demand drops to zero and this, the distance over which consumers are prepared to travel for a specific item, is termed the market area or 'range' of a good. Equally, a certain minimum level of demand must exist before the goods are made available, though this level, 'the threshold' varies from item to item. Expensive and infrequently purchased wares for which shoppers are prepared to travel long distances – such as jewellery or furniture – have higher thresholds and ranges than inexpensive, everyday purchases like groceries and meat. Provided the range exceeds the threshold and the good can thus be made available at a profit, Christaller's central place theory predicts that, in any given market, there will be a large number of purveyors of 'low order' (low threshold and range) goods and relatively few sellers of 'high order' (high threshold and range) merchandise. In addition, as the model stipulates identical sellers, simultaneous free entry and that every customer is served, it follows that the retailers of each item are evenly spaced in triangular arrangement with equally sized, hexagonal market areas or hinterlands, the extent of which reflect the order of the good. When the market areas of high, and every other order of good are superimposed, the famous 'hierarchy' of centres ensues.

Central place theory thus explains the concentrations of high order goods in central places, i.e. nodal points or town centres (central business areas or CBDs in American terminology) of the hierarchy. Brian Berry, one of the leading figures in retail geography in the 1960s and 1970s, put it as follows in his famous study of Chicago in 1963: 'Only the CBD . . . has a large enough trade area to support these (high order) functions', and again 'The

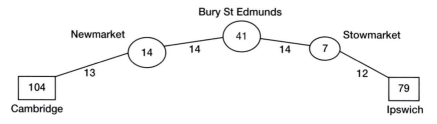

Figure 5.1 Central places in East Anglia
Numbers inside boxes and circles are the 1995 Multiple Branch Score
indicating the size of retailing in the town centre. The numbers between
towns show mileage.

business centre providing good *n* (the good with highest threshold of all)
will require the largest market area for its support, and will presumably be
performed for the entire city by the central business district.' By 1967,
however, Berry showed awareness of the inadequacy of the conventional
theory. He predicted growing specialisation among centres of the same level
of hierarchy and concluded: 'The Christaller type hierarchy thus breaks
down.' In fact Berry wrote at a time when central place retailing in
American cities was rapidly contracting in favour of out-of-town. In the
20-year period from the late 1950s to the late 1970s American CBD
retailing collapsed. In Atlanta, for example, the CBD took 26.4 per cent of
all sales in 1958, but only 7.4 per cent in 1972 (see Urbed *et al.* 1994).

The reason for the failure of central place theory is another of the major
themes of this book, namely the profound locational effect of mass car
ownership. This occurred during the 1950s and 1960s in America and about
a decade later in Europe. The effect was to turn accessibility on its head so
that the town centre fairly suddenly changed from being the most accessible
point to one made inaccessible through congestion. The result has been a
huge growth of dispersed retailing in the last quarter of the century which is
described in the following chapters.

Even in the days before mass car ownership, central place theory had
many critics because of its determinism and the number of heroic
assumptions on which it was based. Where there is a flat agricultural plane,
however, as might exist in East Anglia, it is possible to see settlement
patterns which approximate to the theory. The A14, the main east–west
trunk road running from Birmingham to the docks at Felixstowe, for
example, passes from Ipswich to Cambridge via Bury St Edmunds, as does
the railway. The two regional centres thus have a smaller centre midway
between them, as the theory would suggest. Also midway between Bury
St Edmunds and Ipswich lies the smaller centre of Stowmarket, while
Newmarket performs a similar role between Bury St Edmunds and
Cambridge, again as the theory would suggest.

The theory is still of value in understanding how patterns of service centres developed in predominantly flat agricultural areas, but it is no longer relevant and is indeed positively dangerous as a guide to planning policy or in understanding retail location as it exists today.

Gravity models

Whereas central place theory and many other spatial models are German in origin, the gravity model and related theory on retail location are American. To explain the idea of the gravity model it is necessary to return to the analogy of the earth and the sun with which this chapter opened. The gravity model is based on the idea that the shopper is attracted to a cluster of shops by their mass and deterred from visiting them by their distance. The deterrent effect of distance is more powerful than the attractiveness of mass and is raised to a power in the model. The idea is very similar to Newton's idea of gravity as applied to the planets. In human terms mass becomes attractiveness and distance some hindrance to enjoying that attraction such as mileage, time or cost. The concept is certainly appealing because of its simplicity. If human interaction can be predicted in terms of only two variables this gives great insight into a seemingly complex matter.

In fact there is evidence that a gravity model can give a reasonable estimation of the actual data for a wide range of social phenomena. It was found applicable to railway passenger flows in the US in 1933 by Zipf (1949) and to airline passengers and telephone traffic in America by Hammer and Ikle (1957). Stewart applied this idea to a whole range of social phenomena such as the addresses of college graduate students relative to their college. It has been applied to freight flows and shopping by many such as Huff (1963), also to migration by Hagerstrand (1957) and to residential development by Lowry (1964). The success of this widespread application is striking, particularly when applied to large aggregations of data. The weakness of the concept is the obverse of its strength. It is too simple and generalised. Also human beings are not iron filings responding to magnets but motivated by a whole complex web of forces.

The shopping gravity model broadly assumes that a shopper's likelihood of choosing one supermarket or shopping centre in preference to another is in proportion to its relative attractiveness (mass) and in inverse proportion to its distance, the distance being raised to a power. It is common practice to find where shoppers actually go for their shopping by interview survey and then try to fit a mathematical gravity model to the results. The advantage of this is that it enables the effect of a new shopping centre or supermarket to be measured and predicted. The problem with this is that it is not obvious what proxies should be used for mass and distance. In the case of mass or attraction it is common to use retail sales or floor space, though these have to be estimated as there are no published figures at local level. In practice the shopper may be attracted by other factors such as cheap prices or late openings. Similarly in terms of distance it is common to use either mileage or driving time, but congestion is a factor of growing importance which it is hard to measure as it varies by time of day and day of the week.

Interestingly, American commentators have generally expected the deterrent effect of distance to decline over time as the population becomes more mobile and has more leisure. Similarly the power of distance might fall because of increasing use of the telephone, mail order or the Internet in ordering goods. Whether or not congestion which works in the opposite direction proves to be more powerful, it illustrates the difficulty of using the gravity model. To give a broad brush approximation of human behaviour, the gravity model is invaluable, but as a precision tool it has serious problems.

There are cases in the last century of the idea of gravity being mentioned in connection with human location, but the founding father is recognised as W. J. Reilly of Texas. Reilly used the gravity idea to define the 'breaking point' in the market area of two towns. Reilly's Law as it became known, was first published in 1929 and was used in 1964 in the first part of the Haydock study. This was a report which sought to estimate the impact of a proposed out-of-town regional shopping centre between Liverpool and Manchester and was the first attempt to apply American store location techniques in Britain.

The next major development was to extend the gravitational concept to a range of origins (residential zones) and a range of destinations (super-markets and shopping centres) and model the flow from each origin to each destination. This involved the introduction of the idea of probability in place of Reilly's single break point. The new more complex models produced a percentage for the proportion of the spending of each residential zone which might be expected at each shopping designation. This offered an advance from the determinism of Reilly's Law and made the model more attuned to reality where one person may visit several centres for the same item and where market areas overlap. This advance was mainly pioneered by Huff and later Lakshmanan and Hansen in the mid-1960s.

The 1960s was the great age of urban and regional modelling. Models of the entire urban system were produced then which covered not only shopping flows but also the relationship between different land uses and the traffic flows they generated. In Britain they gave rise to a retail study of the West Midlands and the South Bedfordshire regional study, both published in 1967. Other refinements of the gravity model worth touching on are the ideas of intervening opportunity used in Britain by Black, and maximum entropy developed by Alan Wilson who borrowed the concept of entropy from physics to put alongside the borrowing of gravity. These ideas and further details of the technical side of gravity models are best set out by Ross Davies, formerly director of the Oxford Institute of Retail Management (see Bibliography).

The gravity model in its wider urban form covering a sub-region divided into residential zones and centres has a number of technical difficulties which affect its working. One is that the results are affected by the number of destinations lying outside the study area and by the number of destinations which are included in total. It is difficult technically to close the system. Another problem is that gravity models tend to be fitted to survey data every time they are applied. This can mean that the weight given to distance in a model in one sub-region may turn out to be different to a model of a similar neighbouring sub-region, whereas common sense might suggest that it is unlikely that the inhabitants of these two neighbouring areas behave differently in their reaction to the deterrent effect of distance. These and other problems are discussed fully in earlier works by the author.

We began this chapter by describing the two underlying forces at work in location, the opposing tendencies to cluster and disperse. The gravity model helps in our understanding of these two forces. Mass or attractiveness represents the drawing power that comes from clustering, the synergy resulting from shops locating next to each other. Distance (or time or cost) represents the hindrance that the shopper needs to overcome to enjoy that attractiveness. Substituting time and cost for distance has become increasingly relevant in recent years. Social changes have resulted in people being divided between those who are 'cash rich but time poor' and those who are 'cash poor but time rich'. For the first group time is of great importance and the use of drive time in the gravity model instead of distance would be suitable, particularly if it could allow for congestion, though in practice modellers find it hard to find data which allows for this. For those who are cash poor but time rich attraction will be strongly influenced by price.

The gravity model is therefore a useful tool for the analyst. The basic concept is so simple that it is capable of adaptation to changing circumstances, and, unlike central place theory, it is able to cope with the rise of the car. Its strength is that it specifically recognises the conflict between the two opposing forces of clustering and dispersal.

Retail location models

There are two families of models used in retail location according to the purpose required. For planners and administrators central place theory has met the need to provide a logical framework for the allocation of new population and the provision of the services needed. Thus it was used by the Dutch government in laying out the settlement pattern of the new area of land produced by draining the Zuider Zee. A completely different approach was adopted in America to meet the demand by retailers and shopping centre developers in planning out-of-town stores and shopping centres. The pioneers were all American as might be expected in the nation which pioneered out-of-town shopping. In addition to Reilly and Huff with the gravity model, the leading figures were Nelson and Applebaum, followed by

Cohen and Larry Smith and Gruen who were able to support their ideas with a wealth of practical experience.

There are two main techniques, the vacuum or residual method and the market share or analogue method. These are set out by Applebaum (1965) and in Nelson's standard work on the subject (1958) and covered in Britain by Jones (1969) and in more detail by Ross Davies (1976). The vacuum method seeks to find a gap in retail provisions. It estimates the sales available from the catchment population, usually allowing these to grow in future with population increase and growing personal spending. This is then compared with the sales required by the existing shops. Again allowance is made for this to grow and for new developments which may be in the pipeline. The 'vacuum' is what is left over for additional shopping. It is a method often used by consultants working for local authorities seeking to estimate the impact of a proposed new development on existing shops.

The market share or analogue method seeks to estimate the share of the available expenditure which the new development might expect to capture. The first step is to define a trade area, usually divided into two or three parts. The primary catchment area is that area within which the new store is likely to be the dominant force and within which its penetration rate is likely to be over 50 per cent. This inner area usually accounts for over half of the new development's sales. There is then a secondary and maybe a tertiary area over which the new development will have a significant though minority pull.

The drawing of these trade areas and the estimation of the share of trade which might be captured may be derived from field interview survey, gravity models or experience from other similar developments elsewhere. This method is particularly useful for a multiple retailer, such as a supermarket chain which has a wealth of experience to bring to bear.

In such a case the retailer can also use multiple regression. This is particularly useful for a retailer with a large number of branches, such as a shoe shop, which trades within a shopping centre or major town centre. In this technique all the factors which might influence sales are measured. These might include floor space, competition, pedestrian flow, availability of parking and so forth. These data are then fed into the computer and the model produces the average or expected sales for each branch which can then be compared with the actual sales. The technique has two valuable uses. First, it provides a way of measuring the performance of individual branches. If a branch does badly given its circumstances this might suggest that it is badly managed. Second, the technique is a useful way of estimating the likely sales of a possible new branch.

The technique used depends heavily on whether the retail client likes to cluster together with its competition or prefers to be dispersed. Not only durable multiple retailers like shoe shops seek to cluster together, but often shopping centre developers as well. In America at key freeway interchanges in major metropolitan areas, there are often groupings of shopping centres.

On these occasions it is harder to draw the normal primary catchment area over which an individual shopping centre can claim dominance. The drawing of trade areas may still be undertaken in this case, but the more trade areas are seen to overlap, the harder it is to apply traditional trade area analysis, and the further the analyst is pushed in the direction of using market share methods and multiple regression.

Nelson was also interested in why certain types of retailers were prepared to pay a premium of high rent and other costs to locate near to similar retailers, even competition. His idea was that clustering may generate additional trade to both retailers where two are located side by side. He derived an equation to express this and a rule which states:

> If there are two retail stores side by side and one customer in a hundred makes a purchase in both, then together they will do 1 per cent more business than if separated by such a distance as to make this interchange impossible or unlikely.

Nelson's equation distinguishes between the larger and the smaller store, the inference being that it is the smaller store which benefits most from its big neighbour. This certainly underlies the assumption widely used in shopping centres where the developer provides the anchor stores with attractive financial terms and recoups through the high rent charged to the smaller shops. These shops are prepared to pay high rents because they believe they benefit from the shoppers attracted to these anchors or magnet stores.

Nelson distinguishes between magnet stores which he calls 'generative' and small dependent shops which he calls 'suscipient'. The former are usually large and specialised enough to attract shoppers to make a separate journey to them. In other words they generate traffic and become a destination on their own. The suscipient shop in contrast is dependent on custom which is principally there for another purpose. The station bookstall is an obvious example. Nelson argues that the relative mix of these two factors affects the freedom the shop can exercise in its location and so the rent it need pay. He goes on to develop this into a function which he calls the 'Rule of Retail Compatibility'.

There is a notable difference here, apart from just the size of shop, between different types of durable retailing. Some specialists, like hobby shops, though small in themselves gain little from surrounding retailers. They are destinations, though probably not anchors and as a result can choose dispersed, low rent locations. Others, notably c
shops and shops selling jewellery and household goo(
comparison and so benefit from being part of a clust(
offers a wide choice. In return they are prepared to p
these benefits.

In the following chapter we examine the working (
theory and practice. This difference between comp

retailing is of great importance because the specialist retailer, which according to central place theory should stand in the hierarchy at least as high if not higher than a comparison store, in fact does not necessarily cluster into expensive centres at all.

6 Hierarchy

Hierarchy and ranking

The concept of hierarchy involves ranking towns or shopping centres into grades or ranks according to size or function. It is useful to be able to do this, to talk of village, town and city, whether or not there is an underpinning of theory to justify it. The idea of the hierarchy requires there to be fairly clear differences between different ranks so that it is not too difficult to tell what type of animal we are looking at. The alternative to the hierarchy is the continuum where the settlements rise continuously in size of population. Here the concept is of a continuous line from a large number of small settlements to a small number of large cities, with the level of service provision rising pro rata or linearly. The hierarchy in contrast requires there to be breaks in this continuum.

Which of these concepts is right has long been the subject of academic debate, well described by Haggett. Certainly from observation if there are breaks in the continuum they are not easy to see, and it is necessary to dig a little. The main theoretical support for the continuum idea comes from the rank size rule. This states that the population of a town can be predicted by knowing its size ranking. Thus the 10th largest town in the country will be a 10th the size of the largest city, the 20th will be a 20th and so on. This relationship was first noted in 1913 by Auerbach and explored further by Stewart (1958) and Berry (1961). For the hierarchy the main tradition is central place theory which was discussed in the previous chapter. Central place theory led to attempts to measure 'centrality' through deriving various measures or indices, many using non retail functions such as the presence or absence of theatres, hospitals, colleges, evening newspapers or league football teams. Well known British examples are the work of Smailes (1944), Carruthers (1957), Thorpe (1968), and Smith (1968). Centres were graded in various ways often with natural breaks in the frequency distribution being sought to justify breaks between grades and the hierarchy.

American retail location methodology follows the pioneering work of Applebaum (1957) and Nelson (1958). It tends to regard retailing as more

Table 6.1 Retail hierarchy

Grade	Examples	Multiple branch score	Number of centres	Cumulative
I. National	London	100+	1	1
II. Metropolitan	Nottingham	65–99	12	13
III. A. Major regional		30–64	82	95
	Norwich	50–64	15	28
	Derby	40–49	19	47
	Maidstone	35–39	21	68
	Wakefield	30–34	27	95
III. B. Minor regional		16–29	106	201
	Poole	25–29	30	125
	Lancaster	20–24	39	164
	Folkestone	16–19	37	201
IV. A. Major district		5–15	284	485
	Kendal	11–15	70	271
	Penzance	10	12	283
	Huntingdon	9	19	302
	Exmouth	8	27	329
	Grantham	7	29	358
	Tewkesbury	6	54	412
	Skipton	5	73	485
IV. B. Minor district		2–4	413	899
	Ripon	4	110	595
	Warwick	3	124	719
	Glastonbury	2	180	899

Source: Schiller and Jarrett (1985)

Note
Central London has 29 centres of which 4 are in hierarchy, namely, Oxford Street area, Knightsbridge, Kensington, Chelsea.

an independent marketing function and less a facet of centrality. Whereas central place theory has its origin in the old and dense settlement pattern of Germany, the study of retail location grew among the free-standing out-of-town retail development of low density America. Its emphasis is more on function and shopper behaviour. Both traditions rank and classify centres.

In Britain there was a Census of Distribution in 1950, 1957, 1961, 1966 and 1971. In 1961 and 1971 data covered the larger central areas so giving an indication of centre size in terms of retail sales. Since 1971 there has been no generally available sales data to keep the series going and various commercial firms have published rankings in attempts to fill the gap. In 1985 the author published a ranking using the number of national durable multiples present in each town centre to give a score (Schiller and Jarrett 1985). This in turn was used to suggest a hierarchy as shown in Table 6.1. Although the hierarchy contains four grades, this is to allow for the unique

Table 6.2 The functional division between regional and district centres: the distribution of stores through the hierarchy

Multiple branch score	Proportion of centres with		All stores: number of branches		
	M&S (%)	Other stores (%)	2+ (%)	1 (%)	Nil (%)
20	86	71	57	43	–
19	91	82	73	27	–
18	67	67	67	33	–
17	100	67	67	33	–
16	88	75	75	13	12
15	33	44	11	67	22
14	25	38	13	50	37
13	60	40	25	50	25
12	50	43	21	57	21
11	63	37	26	42	32
10	42	58	25	50	25
9	16	26	5	37	58
8	22	15	–	33	67
7	17	10	–	28	72

Source: Schiller and Jarrett (1985)

position of London and to distinguish the metropolitan cities such as Birmingham and Manchester from the majority of regional centres. The key difference is between regional and district, and the division between them is functional.

District centres satisfy the majority of the weekly shopping needs of their catchment population. They offer both food and other types of convenience shopping as well as the more routine type of durables such as basic clothing. Although they may satisfy the majority of shoppers' durable needs, the level of choice provided is limited. Regional centres in addition to performing the district centre function offer a much fuller range of durables or comparison shopping, enough to entice shoppers from nearby district centres who are looking for, say, a wedding present where wide choice is important.

It was found that the difference between the two grades lay in the presence of stores. Stores are large shops with a sales area of at least 10,000 ft^2 (1,000 m^2). The standard shop unit by contrast is only a tenth of that size with a sales area of little more than 1,000 ft^2 (100 m^2). Furthermore it was found that there was a natural gap in the distribution of store numbers either side of a multiple branch score of 16. Above that there tended to be several stores, below it not even one. This was an interesting finding and it is worth illustrating in Table 6.2. In Britain, stores divide between department stores such as Debenhams and House of Fraser and variety stores such as Marks & Spencer and Bhs. This is because compared with other countries, particularly Germany and America, department stores are weak in Britain and their role has been taken by variety stores. Marks & Spencer had around 300 branches while C&A Modes, Littlewoods and Bhs had around 100 each, while the two big department store groups, House of Fraser and Debenhams each had between 40 and 100 branches. This suggested that there would be one break in store numbers at around the top hundred town centres, and another where the Marks & Spencer stores stop around the 250–300 mark. In fact the two breaks coincided showing that towns tended to have either several stores or none. In other words the break between regional and district centres was surprisingly clear.

The ranking exercise of 1985 was repeated in 1991 and 1996, the former being described by Reynolds and the latter in a report by Hillier Parker. The result has been to establish a time series of change in retail rankings of town centres. One of the results to emerge was the contraction of the number of regional centres from the 201 in 1985 to 172 a decade later. This appears to be a continuous process as the greater mobility given by car ownership satisfies an increasing desire for choice. The last 20 years has also seen the emergence of out-of-town regional centres such as the Metro Centre near Newcastle and Merry Hill in the Black Country. These have been added to the list of centres so concealing to some extent the degree of contraction in the number of regional town centres (see Teale 1997).

Among the other long-term changes to emerge have been, as might be expected, decline in rankings by centres in the North and Inner London where there has also been population decline, and rises in the growth areas around London. Where centres had been split (see chapter 4), there has been a tendency for the stronger to prosper at the expense of the weaker part, and in regions without a clear regional capital again the strongest candidate has prospered. Thus in the East Midlands, Nottingham has performed better than Derby or Leicester and in Yorkshire Leeds has emerged head and shoulders above the previous competition from Bradford and Sheffield. Another interesting result was the rapid rise in ranking of historic or tourist towns such as Chester, York and Bath. The 1995 ranking (published in 1996) showed these towns, together with Oxford and Cambridge, in the top 30 in the country and alongside cities of several times their populations.

Finally, town centre rankings were seen to be highly sensitive to redevelopment. Towns which had transformed their centres by new shopping schemes such as Watford and Bromley had risen rapidly in rank. In fact almost all the top 150 town centres have experienced redevelopment in the last 25 years and this more than any other single factor has determined their performance in the national rankings. Indeed it is the concentration of new development in these top 150 towns which has led to the contraction in the number of regional centres, the trend to fewer but bigger which runs as one of the threads through this book.

The traditional retail hierarchy

An important point about the concept of hierarchy is that competition is as much vertical as horizontal. This means that levels of the hierarchy compete with those above and below (vertical) as much if not more than they do with similar levels (horizontal), and this applies particularly in the retail area. Take, for example, a family living on the edge of a small town of 35,000 population. For food and other routine shopping they will probably spend most at the nearest supermarket, but they could also shop at a close by convenience store, or at a larger superstore on the edge of the large town of 150,000 population which they occasionally visit for clothes shopping. In this case the competition for food is between the local, district and regional levels of a single spatial hierarchy. They are less likely to visit a supermarket in the neighbouring town of similar size, although this may well happen, particularly if they work there or have family links.

In central place theory this interaction of the grades of a hierarchy is called nesting although the competition between the grades is ignored. Central place theory predates widespread car ownership which is what makes this competition possible. Competition between grades is also harder for the researcher to predict. Car use has the effect of making catchment areas overlap and by measuring the effect of distance (or driving time) on shoppers' decisions it is possible to model how expenditure is likely to be allocated between competing supermarkets. If our example family lived between two supermarkets it would be easier to predict which one they would choose, using gravitational and other methods, than it is to estimate the split of their expenditure between convenience store, supermarket and distant superstore. Vertical competition, in other words, is harder to handle than horizontal competition.

Retail hierarchies have been transformed in recent decades since the arrival of widespread car usage for shopping and it is tempting to dismiss the whole concept of hierarchy as obsolete. The example just given shows that this is not the case, although the concept of hierarchy is only one of a number of locational factors and far less important than it once was thought to be. In this chapter we start by looking at the traditional retail hierarchy and then, in the next section, try to develop a replacement which might serve for the present day.

At the bottom comes the free-standing village shop or corner shop. This traditionally sold food, but because of the strange way newspapers were distributed in Britain, until quite recently it often could not sell newspapers. This led to the existence of the confectioner, tobacconist and newsagent, known by various names to the public and known in the trade as the CTN shop. These two types of shop were largely superseded by the arrival of the supermarket. With 1980 equalling 100, the number of grocers fell from an index figure of 299 in 1961 to 78 in 1984, a drop of nearly 4:1 in only 23 years. CTNs fell 36 per cent in the same period. Those that survive today perform a supportive, top-up roll, often staying 'open all hours'. The bottom level of the hierarchy is filled in addition by the shop within the petrol filling station and the relatively few American-style convenience stores.

Next came the neighbourhood centre or small parade of between say five and 50 shops. These were planned into the first generation of post-war new towns and had difficulties from the start. With the total number of shops in the country roughly halving since 1961, this type of centre has been widely predicted by various government reports as being likely to follow the corner shop into history. In fact many survive in full occupation, but the role they play has changed and is no longer a traditional service function in the hierarchy. Today, instead of providing convenience shopping for their neighbourhood with butcher, greengrocer, baker, sub-post office and chemist to support the CTN and grocer they offer mostly services and non-retail functions. A parade of a dozen shops today might include a hairdresser, a laundrette, a pizza take-away, an Indian restaurant and take-away together with non-retail functions such as auto accessories, central heating and double-glazing firms. Interestingly, too, they often have one or two shops occupied by specialist retailers.

Hatfield Road, St Albans near the author's home, is a typical suburban radial road a mile from the city centre occupied by a continuous straggle of shops of this type. Among the ever increasing number of take-aways and the other uses just described are to be found a shop hiring out crockery and cutlery for parties, a naval and military tailor and a shop specialising in equipment for the disabled. Not far away is a shop offering equipment for the golfer. The significance of this is discussed below, but the effect is to help keep the shop property occupied.

This story suggests that there is generally a shortage of premises available to the non-retail business serving the general public. These businesses which are myriad and growing in number have little choice but to occupy shop property in fringe locations. The only alternative is the industrial estate which is worse in that there is usually insufficient parking for the workforce never mind the general public, and no main road presence to attract passing trade.

The third step in the traditional retail hierarchy is the district centre. This might serve a catchment population of 20,000 to 100,000 and offer a whole range of convenience and routine durable shopping. Its function is to satisfy the main weekly family shop, although until the arrival of the supermarket many families did this in the neighbourhood centre, using the district centre for less frequent trips. The district centre contained banks and building societies and a branch of Woolworths and Boots. Since there are today 1,000 each of these two bellwether national multiples, and since there are around 200 regional centres in Great Britain, it follows that there are around 800 district centres in the country situated in town or suburban centres. District centres might once have contained a branch of Montague Burton, but generally they are too small to attract nationally known multiple durable retailers like Next, H Samuel or WH Smith. They have plenty of durable shops but these are run by independents or regional multiples.

The fourth step in the hierarchy is the regional centre which offers a full range of durable shopping, including the well known national names. Indeed so dominant are these traders that there is a sameness to shopping in major British town centres from Aberdeen to Plymouth which is often commented on. As was mentioned in the previous section, the key difference between regional and district grades lies in the presence of department and variety stores.

In the planning literature these centres are commonly described as sub-regional because their catchment areas are less than a standard region in extent and allowance is made for the inclusion of a grade of large city above them. In the ranking described above these grades were called metropolitan and national but other names are often used such as major regional for the top eight to ten cities in the country. In retailing terms the main difference between a regional centre such as Northampton and a larger centre such as Birmingham is that the latter has more specialist shops and more and bigger department stores. In terms of shopping surveys, the results show that Birmingham has only a weak pull over its region beyond the West Midlands conurbation. The concept of hierarchy is further weakened by the arrival of out-of-town free-standing regional centres, and today a shopper in Northampton seeking a greater choice than was available there is as likely to visit London, Milton Keynes or the out-of-town regional centres at Meadowhall (near Sheffield) or Merry Hill (near Dudley) as Birmingham.

If the hierarchical concept fails above the regional centre it also fails below the district centre. Local shops and neighbourhood centres may appear in size terms to continue to offer a local service, but this is limited to a fairly modest minority of top-up spending. If the population in a small town or village leaves that town for three-quarters of its food shopping and uses the neighbourhood centre only for top-up shopping, it is difficult to argue that that centre continues to function as a distinct hierarchical layer as it did 30 years ago when the majority of food shopping would have taken place locally.

According to the traditional retail hierarchy, Northampton would nest within the regional catchment area of Birmingham which in turn nests within the national pull of London. Beneath would be district centres such as Wellingborough and neighbourhood centres like Finedon. It is an attractive theory, even elegant in its operation, as Figure 6.1 shows. Unfortunately, even in its heyday before the car it did not function particularly well, and today its relevance is limited.

A retail hierarchy for 2000

In looking for a structure to the various forms of retailing at the turn of the millennium it is still just possible to use the concept of hierarchy. The last

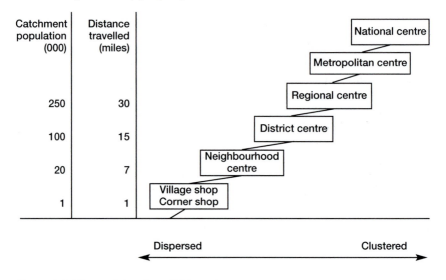

Catchment population (000)	Distance travelled (miles)	
250	30	
100	15	
20	7	
1	1	

National centre
Metropolitan centre
Regional centre
District centre
Neighbourhood centre
Village shop
Corner shop

Dispersed   Clustered

Figure 6.1 The traditional retail hierarchy

decades of the twentieth century have seen the weakening of many of the traditional boundaries in retailing. Take for example the division between convenience and durables. Convenience goods (broadly speaking food and CTN) were bought frequently and locally. Distance was a powerful deterrent because of the weight of the goods involved, and many people did the bulk of their convenience shopping in corner shops or neighbourhood centres. Durable goods, in contrast, were very different. This involved choice and comparison, shopping was undertaken rarely and involved much higher levels of expenditure. Durable shopping was concentrated in the few big centres where big named stores were to be found and usually it involved a different trip which included no food shopping.

This division still survives in part, but the division is nowhere near as clear cut as it was. Marks & Spencer, for example, the nation's leading clothes retailer obtains between a third and a half of its sales from food, while the major supermarket chains, particularly Tesco and Asda, sell increasing quantities of clothing and other non-foods. Government figures on retailing (Census of Distribution, Retail Inquiry, Business Monitor SD25, etc.) used to make a fundamental division between convenience and durable shops, but in recent years have bowed to the inevitable and introduced a category called Large Mixed Business which blurs this division. The purpose of this example is to show that what follows is an examination of the various types of retailing as much as a clear hierarchy. Indeed it is argued that the only hierarchical element remaining is the division between district and regional centres, and even here the division is less clear than it once was.

Let us start with food shopping, the lowest tier of shopping. As we saw in the previous section, corner shops and neighbourhood parades survive in reduced form but they no longer satisfy the bulk of their customers' food shopping needs. There are five main types of shop below the level of the supermarket in terms of size (a supermarket being defined as a self-service food store with a sales area of at least 10,000 ft² (1,000 m²) and these are the corner or village shop, the street market, the garage shop, the convenience store and the small supermarket (sometimes called a mini-market or superette and usually occupied by a voluntary group such as Spar). Another form of small supermarket are the discount stores such as Aldi and Netto.

The two new forms (the petrol filling station shop and the convenience store) are designed to combine various strands of local retailing. The convenience store offers cosmetics and sometimes a simple café or take-away while the petrol station combines sales of petrol with basic retailing. All four types combine food with CTN and all, except possibly the traditional street market, rely heavily on casual or passing car-borne trade. It would be wrong to describe them as sitting below the supermarket in a hierarchical sense because they depend so heavily on top-up trade and therefore draw lightly over a wide catchment area.

There are perhaps elements of irony here. The village and corner shop beloved by conservationists is able to survive only because it behaves like an out-of-town retailer by offering easy car access and parking, and though the concept of convenience shopping has been largely abandoned by statisticians, these shops are in practice heavily dependent upon offering convenience to their customers rather than just proximity.

It is difficult to illustrate food shopping in diagrammatic form, but Figure 6.2 offers an interpretation which shows the supermarket at the centre surrounded by satellites which are joined to the supermarket to show they are not subject to measurement on the axes of distance or clustering. It is interesting to note how far up the traditional hierarchy the supermarket now sits. Although the minimum size of supermarket as defined here is 10,000 ft² (1,000 m²), most are two or three times that size and food-anchored superstores and hypermarkets exist of over 100,000 ft² (10,000 m²). The average supermarket of say 25,000 ft² (2,500 m²) therefore in terms of sales is equal to the total of a small town centre, while it is difficult to see the larger superstore or hypermarket with sales in the £50–£100m range as other than matching the facilities of a substantial county town centre.

As we move on to consider non-food shopping we are not necessarily moving to bigger concentrations. Large supermarkets or superstores (the division between the two is often made at 25,000 ft² (2,500 m²) which is unfortunate as many modern supermarkets are in the 20,000–30,000 ft² (2,000–3,000 m²) range) and are sometimes built as the anchor to a shopping centre which is commonly called a district centre. As we saw in the previous section the district centre is liked by planners because it fits into

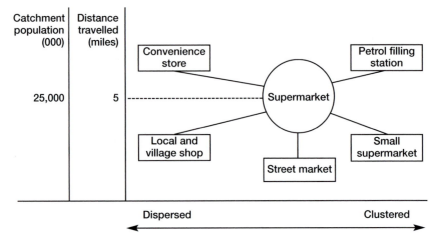

Figure 6.2 Food shopping

the traditional hierarchy. It is a little confusing because the term district centre or regional centre can refer to either a town centre of the appropriate size or function or an out-of-town shopping centre, and sometimes to a purpose-built shopping centre within a town centre. The important point to look for is whether the town centre or shopping centre is anchored by food or durable shopping. By the end of 1998 there were 1,300 shopping centres in the UK, half of which were built to augment existing town centres. The majority of these are predominantly occupied by durable retailers and were built to expand and modernise the durable shopping provision of the towns where they are situated, usually forming the prime area with the highest rents. As we saw earlier there has been a tendency to concentrate the regional centre function in the top 200 towns, supplemented by a small number of no more than a dozen new out-of-town regional centres which function as additions to the list. The larger of these centres are strong enough because of the high level of clustering to resist the competition of dispersed forms of durable shopping, and in that sense the out-of-town regional centres such as Brent Cross or Bluewater are additions to a list which includes strong regional town centres such as Harrow or Bromley, rather than competition to it.

The regional centre forms the central function of durable shopping in a way similar to the supermarket in food shopping. There are in addition various specialist types of development such as factory outlet centres and these are described in a typology in the next section. The only one of these which needs mentioning here is the retail park which accounts for 35 per cent of all shopping centres (around 500 centres) and has grown very rapidly during the last 15 years of the century. Britain is unique among advanced countries in having such a large number of shopping centres built

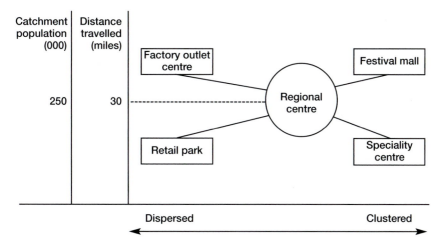

Figure 6.3 Durable shopping

within town centres and also in the number of retail parks. These are almost all outside town centres and consist of a handful of retail warehouses of 10,000–40,000 ft² (1,000–4,000 m²). The backbone of retail parks is DIY, furniture and carpets and discount electrical goods. Increasingly in recent years general durable retailers have also begun to open in them because planning opposition has limited the opportunity in more conventional forms of out-of-town shopping centre. Many retail warehouses are found in free-standing locations rather than in parks and there has been a tendency for these to group themselves in ad hoc concentrations along stretches of main road. London examples of this are Purley Way, Croydon and Edgware Road between Cricklewood and Burnt Oak. Valuation evidence suggests that the retail parks and natural concentrations attract higher value than free-standing retail warehouses. Rents are higher and yields are lower for retail parks and natural groupings than for free-standing locations. There seems to be a slight benefit for retail warehouses to cluster together. This makes it all the more strange that Marks & Spencer when it decided to decentralise opened the majority of its new out-of-town stores in free-standing stores, commonly accompanied by just a neighbouring superstore. In America and most other countries clothing stores cluster within shopping centres, so it is perhaps an indication of the confidence Marks & Spencer has in its own strength as well as the lack of opportunity there has been for it to anchor out-of-town regional shopping centres, that it should behave in this strange way.

Figure 6.3 shows durable shopping in a way similar to food shopping in Figure 6.2. Most of the specialist types such as festival malls and factory outlet centres have catchment populations and clustering tendencies similar to regional centres so they have been shown attached to the central regional centre form and divorced from the axes. The retail park is also found within the catchment area of regional centres, and few are found in smaller or remote areas. Their distinctive feature is that because they tend to remain small with an average size of 150,000 ft² (1,500 m²) several are found in the suburbs of each of the larger towns and cities. It would be a mistake to conclude that they sit below the regional centre in hierarchical terms as a result. DIY which was the origin of the retail park has a low threshold population and can be found in district centres, but furniture, carpets and discount electrical goods are classic examples of durables. Retail parks, therefore, are generally on the same level of hierarchy as regional centres.

In this section it remains only to say a word about specialist shopping, that type of retailing which demands a catchment population of a million or more and for which shoppers might be prepared to travel for an hour. Retailers of this type include department stores such as John Lewis, Harvey Nichols and Selfridges and specialist retailers such as international designer fashion stores. It also includes esoteric independent retailers such as shops specialising in theatrical costumes, rare stamps, model railways and a host of similar areas. According to central place theory this type of retailing should be limited to the centre of metropolitan cities, but as we saw earlier specialist independents are often in practice dispersed to low value tertiary shopping positions such as Hatfield Road, St Albans. This needs an explanation.

John Lewis have in recent years opened branches in Bristol, Aberdeen, Glasgow and Manchester and announced plans to open near Birmingham. In doing so they have limited themselves to the great cities or metropolitan centres, and appear to confirm the traditional hierarchy. But in only two (Glasgow and Aberdeen) of the five cities listed, however, is the new branch to be found in the city centre. In the remaining centres the new branch serves the catchment area from another location. In Bristol and Manchester it is on an out-of-town site and in Birmingham it is in Solihull, a regional centre in a middle-class suburban area. In these cases it could be argued that the new location is better placed and more central to serve the upmarket customers of John Lewis than is the city centre itself. If the catchment population is becoming increasingly polarised with segregation increasing between levels of income, and there are signs of this happening, then a retailer seeking to serve upmarket customers is torn between the major centre with associated high costs and a smaller centre or out-of-town site which is more accessible to these customers.

If one imagined a professional woman, say a solicitor or accountant, living in Solihull but working in central Birmingham, then a fashion shop such as Jaeger seeking her custom can choose between being accessible to her during the working day in central Birmingham, or at weekends in Solihull. It is not an obvious choice, but many retailers have followed the policy of John Lewis and opted for the lower cost solution.

As long ago as 1971, the author noted in an article (Schiller 1971) that

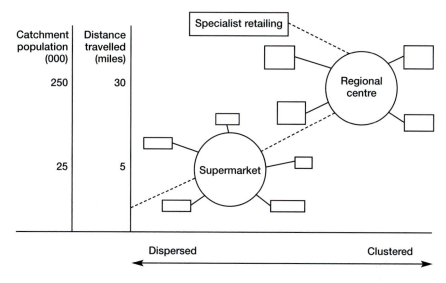

Figure 6.4 Specialist shopping

services in the area surrounding Greater London were not behaving in the way that central place theory predicted, and that upmarket services such as garages selling imported luxury cars, shops accepting the Diners Club credit card, and stockists of products mentioned in *Vogue* magazine, were locating to a disproportionate degree in certain quite small upmarket centres. Thus Beaconsfield contained 27 such outlets compared with only four in Slough, although Slough boasted a conventional shopping centre many times the size of Beaconsfield. Like John Lewis in Solihull or Cheadle near Manchester, these retailers found they could reach their target more efficiently outside the traditional town centre even though they were dependent on a catchment area large in terms of both distance and population.

The same argument applies to the more esoteric independent retailers which were noted in Hatfield Road, St Albans. These retailers attract their customers through advertising and word of mouth rather than passing trade. Provided they are located in an area with an adequate catchment population within a reasonable distance, offer parking and are reasonably easy to find, they are free to locate wherever costs are lowest.

Once again it is the car which has turned central place theory on its head. The city centre is no longer the unique point of maximum accessibility. Figure 6.4 tries to show this in diagrammatic form. It shows that for specialist retailers who need a catchment area larger than a regional centre, the need to cluster is no greater than for a regional centre and can be less. It suggests, too, that attempts to reinforce metropolitan centres to levels larger than a regional centre might well be dangerous and short-sighted.

A typology of shopping centres

In this final section of the chapter on hierarchy we offer a classification of shopping centres. A shopping centre is defined as a development which was built and let as an entity, contains three or more retail units, and includes a purpose-built pedestrian area outside the component shops or common car parking. Most shopping centres are over 50,000 ft² (5,000 m²) in size and that is the definition used here, although a number of specialist centres are smaller than this.

The typology is limited to centres which were being developed in the 1990s and which has the effect of excluding smaller centres built in the 1960s and lower order centres in the earlier new towns such as Hemel Hempstead and Harlow. Shopping centres as defined here are not the same as retail concentration in town centres, although these may include shopping centres. Of the eight types of centre listed, most can be found in town centres although all eight can also be found outside town centres.

District centres

The key feature of district centres is that they are food dominated and anchored by a superstore where the majority of the sales comes from food. There are some 50 of these in out-of-town locations in Britain, examples being Lords Hill in Southampton, Beaumont Shopping Centre in Leicester and Cameron Toll in Edinburgh. In size they range from 50,000 to 250,000 ft² (5,000–25,000 m²) and they first appeared in the 1970s. Apart from the food anchor district centres commonly contain few specialist food shops but would include a chemist, post office, takeaway, hairdresser, newsagents and stationers and basic clothing shops.

The district centre has a threshold population of 50,000–150,000 and is found mainly in the suburbs of large cities as the examples given suggest. They are more rarely found today in the centres of small towns although some shopping centres built earlier would be classified as district centres, for example St Martin's Walk, Dorking and the Quadrant, Dunstable. Despite the large number of supermarkets and superstores (over 1,000) district centres are not common because superstores find it unnecessary to form the anchor of a shopping centre. The majority of superstores are free-standing and offer an increasing number of services and non-food goods within the envelope of their own stores.

Regional centres

Regional shopping centres offer durable or comparison goods shopping. Because this type of shopping benefits by clustering together, they are therefore of substantial size and generally contain large stores. Most durable shopping in Britain takes place in town centres, and most town centres which are large enough to qualify as regional centres (roughly the top 150) have had their centres strengthened and expanded by one or more shopping schemes which vary widely in size. There are 350–400 such schemes in the top 200 centres. On their own these are mainly not sufficient to qualify as regional centres. It is only when joined with the remainder of the town centre that the whole complex provides sufficient choice to qualify.

In America regional shopping centres are taken to have a minimum size of 500,000 ft^2 (50,000 m^2) with a minimum of two department stores. In Britain there were only 25 shopping developments of that size or more opened at the end of 1994 and fairly few (such as Bluewater near Dartford) which have opened between 1994 and the end of the century. These include centres such as Brent Cross with its classic dumbbell layout of a mall running between two anchoring department stores, as well as other out-of-town regional centres such as Lakeside and Merry Hill. They also include centres of new towns like Milton Keynes and Telford and town centre schemes such as the Bentall Centre, Kingston and the Harlequin Centre, Watford. Regional centres contain a full range of durable shops but there is a particular emphasis on fashion clothing and shoes. The minimum threshold population is usually reckoned at 250,000 although there are many centres with a smaller catchment population which offer a wide range of durables.

There is no agreed upper size limit to the regional centre although in practice there are few over 1.5m ft^2 (150,000 m^2). Most lie within the range 750,000–1,250,000 ft^2 (75,000–125,000 m^2). In America where there are larger retail conglomerations they tend to consist of several shopping centres at a freeway intersection such as at Tysons Corner in Washington DC or Post Oak in Houston. Although it is possible to point to giant shopping centres of over 2m ft^2 (200,000 m^2) and above, such as Mall of America and West Edmonton in Canada, their rarity suggests that the shopper gains little by clusters much over 1.25m ft^2 (125,000 m^2). They become too big to walk round and retailers tend to duplicate as a result. The two branches of Marks & Spencer in Oxford Street is an example of this.

Retail parks

Retail parks are collections of retail warehouses usually occupying three sides of an open square facing a main road. Unlike the regional centres which are usually enclosed and air-conditioned, retail parks are open to the elements and resemble the industrial estates from which they evolved. They have grown very rapidly in the last 15 years of the century from 11 in 1985 to some 500 by the year 2000. They are an extraordinary phenomenon, being unique in British shopping history in the great rapidity with which they have grown and the fact that they have no exact counterpart in America, France or most other countries where shopping centres are found.

They began in order to accommodate the rapid rise of DIY which coincided with the decentralisation of furniture and carpets from the town centre during the 1980s. They took advantage of the availability of industrial land following on the contraction of manufacturing and the restructuring of warehousing in this period. Because of this local authorities appear to consider them to be somehow not quite retail, perhaps more industrial and as a result many have limitations forbidding them to contain units of under 5,000 to 10,000 ft² (500 to 1,000 m²) and to sell general retail goods.

Retail parks typically include DIY, furniture, carpets and discount electricals, but in the last ten years pressure for growth has come not from these areas but from a wide area of retailers looking for space outside town centres. Most retail parks are located outside the town centre on former industrial land and have floor areas of 50,000–250,000 ft² (500–2,500 m²). Retail parks tend to be found in the suburbs of towns of regional centre size, a large town having perhaps three or four centres. Although many of their retailers compete with the town centre, because this competition is dispersed, its effect on the town centre is lessened. Retail parks are rarely big enough to form a nucleus of sufficient power to act as an alternative location to the town centre. Another reason is that retail parks tend to include limited clothing and footwear outlets.

Hybrid centres

This category covers those shopping centres which fit into neither the three major categories above nor the four specialist categories which follow. There are two groups of them. The first includes those centres which reinforce the durable role in towns which are too small to be called regional centres. Broadly they are found in towns ranked 200–400 in the national hierarchy. Most of these towns have a shopping centre within their town centre and it would be misleading to define them as either regional or district centres. There are around 150 shopping centres of this type, mostly completed before 1985, but few would be built today.

The second category of hybrid consists of out-of-town durable centres which are too small to reach the minimum size to be called a true regional shopping centre. They are of between 250,000 and 500,000 ft² (25,000–50,000 m²) in size and there are relatively few of them. The Hillier Parker listing of 1995 includes only six and a similar number have been developed since. Examples are The Gyle Centre, Edinburgh, and Hempstead Valley near Gillingham. The very fact that there are so few of them suggests a natural break between regional centres of over 500,000 ft² (50,000 m²), most of which are over over 750,000 ft² (75,000 m²) and other types of centre which are nearly all under half that size. In most cases these centres would grow to be regional centres by market forces if site or planning restrictions did not prevent that happening.

Factory outlet centres

These are centres where branded goods are sold at a discount. In particular they tend to concentrate on fashion brands where unsold lines from the previous season are disposed of. As with most types of shopping centre they are an American import and did not become significant in Britain before 1995 when the total floor space passed the 1m ft² (100,000 m²) level. Floor space then rose rapidly to 4m ft² (400,000 m²) by the end of the century in 30–40 different centres.

There are more of them in Britain than any continental country, France coming second. In size they range from under 50,000 ft² (5,000 m²) to 250,000 ft² (25,000 m²), though many of the American centres are smaller than this. Examples are Cheshire Oaks, Bicester, Clarke's Village in Street, Somerset, the pioneer, and the Great Western Designer Village in the old railway workshops in Swindon. Several have been built around mock village squares and by quaysides in converted dock areas. Unlike regional or district centres they do not rely on magnet stores to attract customers. Catchment areas are large, perhaps an hour's drive, and can run into a population in the millions. Clearly they compete for durable goods with the larger regional town centres, although the impact is often diffused over a wide area. Many planners oppose the spread of these centres on the grounds that they are simply a way round the current opposition to out-of-town shopping.

Power centres

Power centres consist of the largest 'big box' retailers who are 'category killers'. Strictly speaking they should not be included here because they are an American concept which so far does not exist in Britain in precisely this form. 'Big box' is the American term for giant retail warehouse and 'category killers' are retailers who are so powerful in their market that competitors find it hard to survive. Examples in Britain are Toys R Us and Ikea, which offer toys and furniture on a huge scale, more in floor space terms than the total toy or furniture floor space of a large town centre.

The nearest British equivalent to the power centre is the large retail park, such as Team Valley, Gateshead and Parkgate, Rotherham, which are in the 350,000–450,000 ft² (35,000–45,000 m²) size range. Thurrock, at the junction of the A13 and M25 in Essex contains three adjacent retail parks which together total 620,000 ft² (60,000 m²). This retail complex which sits next to Lakeside regional shopping centre contains over 20 furniture and carpet retail warehouses which together have a floor space unmatched in scale anywhere else in Britain, including central London (Figure 6.5).

Although the example given is not a shopping centre in that it was not 'built and let as an entity', it functions in the same way as an American power centre. Category killers and huge conglomerations like Thurrock which are mostly away from traditional town centres could be argued to sit

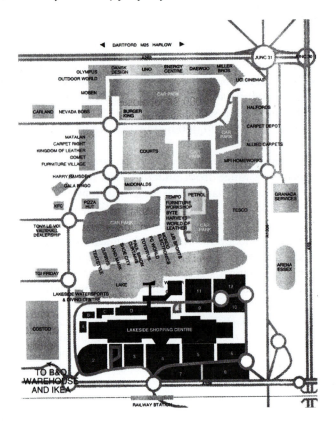

Figure 6.5 Lakeside retail park guide
            Source: Lakeside shopping centre

above the regional centre in hierarchical terms. What they amount to is a clustering of a single aspect of durable retailing to a degree greater than that found in the largest regional centres. The specialist nature of the trip to buy, say, furniture means that it is not necessary for furniture to locate next to clothing or other types of durable retailing.

Festival malls

Covent Garden market is the most successful example of the festival mall in Britain. They are centres which appeal to people engaged in leisure activity rather than the serious business of acquiring goods, and their customers are mainly tourists, office workers and young people. They offer a total experience of which the goods on offer are only a part. Despite its huge appeal, Covent Garden has a retail area of only 52,000 ft^2 (5,000 m^2) of which a sizeable part is cafes and restaurants of various types.

Festival malls were first developed by the Rouse Corporation in America on derelict waterfront sites in Boston and Baltimore. For them to succeed it was necessary for the shopping centre to be combined with other attractions to appeal to visitors. In both Boston and Baltimore the waterfront was improved and opened up and in the Boston case the shopping development was near the historic Faneuil Hall. In the case of Baltimore the equivalent attraction was a giant new aquarium. In London there was no need for a similar attraction because Covent Garden was within a short walk of Leicester Square, the centre of entertainment in the capital.

Other attempts have been made to repeat the success of Covent Garden, for example Albert Dock in Liverpool, but although far from being failures, they have often found it difficult to attract the large numbers of visitors which a festival mall needs to thrive. The festival mall illustrates a new type of leisure trip in which conventional shopping is only a part. The success of retailing in the streets surrounding Covent Garden and the very high rents in historic cities such as Chester, Bath and York is part of the same phenomenon. The leisure shopping trip combines shopping with browsing, visiting and enjoying a pleasant environment and eating out. It is growing rapidly.

Speciality centres

Speciality centres contain small groups of specialist shops selling usually luxury or art goods. Perhaps the best known example is Burlington Arcade in Piccadilly near Bond Street which, dating back to the nineteenth century, could also claim to be the oldest shopping centre in Britain. The best known modern example is Trump Tower in New York. The speciality centre has no anchor store and rarely contains a multiple retailer. It relies on its location in an area known for luxury shopping such as Bond Street or Knightsbridge. It is one of the few types of shopping centre which is found in similar format throughout most of the big cities of the world. Often it is linked to a prestige hotel.

In Britain in the last 15 years a number of centres have opened which appear to be hybrids of festival malls and speciality centres. They are of small size and they appeal to a combination of office workers and visitors (for example the shops attached to the offices at Canary Wharf and at London Bridge City on the Thames near London Bridge Station). Many found it difficult to attract large numbers of visitors and had to adjust to the fact that visitors and office workers are both more likely to spend on eating out than on durables. Those that succeeded have done so with a high level of catering combined with hobby shops and shops catering for basic needs such as chemists and stationery. Outside London there are a number of centres, often in converted historic buildings which contain specialist shops such as antiques. Nearly all are less than 50,000 ft^2 (5,000 m^2) in size and so do not qualify on the definition of shopping centre usually used.

7   The desire to cluster

Minimising the search

So far we have examined the desire to cluster and the desire to disperse in terms of the way shops themselves can be seen to locate, grouped into centres of various sizes or preferring to remain free-standing. In the next two chapters we look in turn at the two underlying forces more from the point of view of the shoppers themselves, and the best place to begin is the search for goods.

There are two distinct steps in the activity of shopping once the decision to buy is made. The first is gaining access to the goods being sold and the second involves making a choice from among them. Gaining access to goods usually, but not always, involves a physical journey. The second step in shopping, which is choosing the goods to buy, is itself in two parts: suitability and price. Suitability is selecting the right quality, style, size, colour, etc., while price is choosing the offer which provides the best value for money. In general it is easiest to think of shoppers as trying to minimise the time, effort, cost and general inconvenience of both travelling and searching to get what they want. It is tempting to slip into economists' jargon and talk of utility and optimisation, but the concept is really quite simple.

Some people enjoy the process of choosing, others window shop for pleasure or buy on impulse, and increasingly shopping takes place as an incidental activity as part of a leisure trip, but this does not affect the basic concept. For the large majority of shopping, people start with a list and try to get everything on their list as painlessly as they can. The amount of effort they are prepared to spend both in travelling and searching depends on the importance they attach to the items on the list. At one extreme come items like cigarettes and newspapers. There is no issue of quality or price here as most people will buy the same brand of cigarettes or newspaper title as previously. In other words the selection has already been made and the shopper does not need to waste time wondering which type to buy. Since both items are branded the quality (and usually the price) is the same wherever it is bought. For purchases of this type the sole consideration is convenience, minimising the effort of travel.

The result is that cigarettes and newspapers tend to be sold in large numbers of small kiosks, which tend to be located near existing pedestrian flows, for example at railway stations. There may be a market for a handful of specialist tobacconists or for newsagents carrying foreign newspapers, but for the large majority of retailers in these fields there is no point in offering a wide choice because their customers already know what they want. They are not destinations but rather offer a service to be used on the shopper's trip to somewhere else. Their best bet is to behave like parasites and attach themselves to activities which attract crowds such as football stadia, hotels, shopping centres, motorway service stations and airports.

At the other extreme come items which require thought, and the best example of this is the wedding present. A wedding present makes a statement about the shopper's relationship with the happy couple. It must not suggest either ostentation or meanness and shouldn't duplicate other gifts. It needs planning and care and saving money and effort is not top of the list of priorities. Someone buying a wedding present could be prepared to spend considerable effort, more than usual on both the journey and the search to get the satisfaction of feeling that a right choice had been made. In between these two extremes come all other items ranging from food through routine clothing to fashion clothing and jewellery. Shoppers vary in what they regard as important and what is considered just a routine chore. Some take an enormous interest in food but care little for clothing whereas others might devote great attention to hi-fi or a hobby and regard both food and clothing as convenience items. Generally, however, the more frequently purchased items such as food and other groceries attract less attention and so are referred to as convenience goods while shoppers are prepared to invest more time and money in the search for less frequently bought durable items which as a result are often called comparison goods.

This variation in the attitude shoppers have to the amount of effort they are prepared to devote to buying various types of goods led to the creation of the hierarchy as described in the previous chapter. It is also a strong argument in favour of clustering. The shopper wants to avoid the disappointment of not finding the precise item wanted when it is something that is cared about. The shopper has already invested heavily in the journey in order to have access to a range of choice so it is important that the choice is maximised by the clustering together of similar retailers.

The ice-cream man

Shops which cluster together in order to provide a wide range of choice also have the effect of offering the benefits of competition such as lower prices. This point is illustrated by the story of the ice-cream man which is attributed to Harold Hotelling half a century ago. Imagine a beach covered with holiday-makers evenly spread over the sand, and in their midst stands Mario, the ice-cream man. It is obvious to him to stand in the middle

(a) Mario and Luigi divide the market

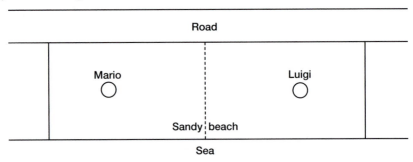

(b) Mario and Luigi compete for the whole market

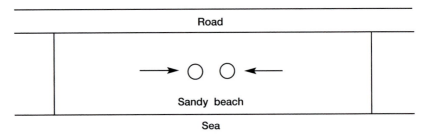

Figure 7.1 The ice-cream man's dilemma

because by doing so he minimises the distance his customers have to walk. The problem arises when his rival Luigi arrives to take up a position. To keep it simple we will assume that both men's equipment is too heavy for them to walk up and down.

Common sense leads Luigi to argue that Mario should move up to divide the beach between them and Mario agrees thinking that the further away Luigi is the better (see Figure 7.1). The market is thus split between them into two equal halves. This continues until Luigi notices that if he moves towards the centre he will increase the size of his market at the expense of his competitor. Mario notices what is happening and responds by moving towards the centre too, and the game continues until Mario and Luigi end up back to back in the centre of the beach. This story shows that from a retailer's point of view it can pay to locate next to a competitor. In doing so they are maximising their access to the market and allowing them to compete on grounds other than distance such as price, range of goods, quality and service with a smile. The case of the ice-cream men makes the point because it is easy and cost free for them to move on the beach, whereas this is not the case with retailing from the fixed premises of a shop.

The story refers to the benefit gained by the retailer rather than the customer. The same logic which brought Luigi and Mario to trade back to back would also apply to a third, fourth or any number of ice-cream men and this is clearly nonsense. In real life we do not find clusters of ice-cream men in the middle of the beach. The reason for this is that the customer has a different set of priorities to the ice-cream man's. Ice-cream is a fairly marginal purchase, a convenience buy, and the amount the holiday makers on the beach choose to spend will vary according to how easy it is to buy. People at either end of the beach will spend less if the only source of ice-cream is in the centre than they would have done when Mario and Luigi were located apart and as a result nearer to them. In other words, the story would work in practice only if the demand for ice-cream were inelastic and Mario and Luigi had a monopoly over supply. The situation would be different if instead of ice-cream the item being sold was one which shoppers cared about or if the holidaymakers at the edge of the beach were connoisseurs of ice-cream. In that case they would welcome the additional choice which came from Luigi and Mario being next to each other rather than apart.

Maximising the benefit

Shopping by car has hugely increased the access shoppers have to concentrations of retailers clustering together, and this has tended to increase the level of clustering itself. Since the movement out of town, however, this does not necessarily mean increasing the use of town centres. There are few areas of Britain, amounting to less than 10 per cent of the population, that are beyond an hour's drive of a major retail centre. Catchment areas overlap to a degree that was unthinkable before the widespread use of the car for shopping. The result generally has been to give big centres an advantage over small. Of course factors other than sheer size are important in the selection of a shopping centre. Freedom from the fear of crime, ease of car parking and quality of the environment are all important, but in general a large centre is more likely to maximise the benefits to the shopper because the likelihood that every item on the shopping list will be ticked is increased.

Figure 7.2 shows the increases in the number of multiples (a proxy for size of centre) by centre type between 1984 and 1995 using the method described previously. It shows a bigger increase among regional centres than district centres. Excluding the few metropolitan centres, the effect of widespread car usage for shopping has been to give advantages to the substantial regional centres against the less specialised district centres. The freedom of choice which the car has provided has allowed greater clustering to take place.

The same was found for the same reasons in a major study of the effect of bypasses on retailing. The classic study by Garrison *et al.* (1959) examined a town which was bypassed allowing through-traffic on its way to a bigger city to avoid the town centre. The results were that the town became more attractive once it was less congested and in consequence more shopping took

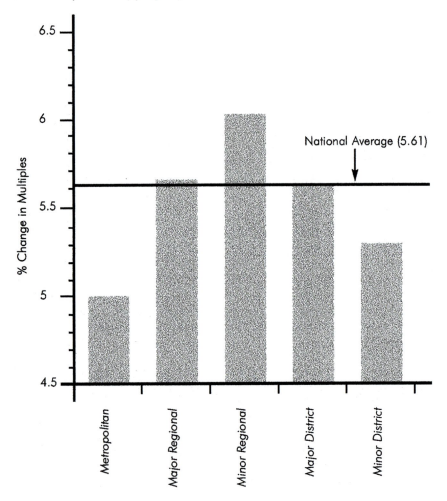

Figure 7.2 Hierarchy: percentage changes in multiples, 1984–1995
Source: CB Hillier Parker

place there at the expense of the small towns nearby. Similarly it was now easier for shoppers in the neighbourhood to reach the big city, and as a result the more specialised types of comparison shopping were found to have switched there from the town which had been bypassed. In other words there had been a general trend for greater access to result in shopping to take place further up the hierarchy than before.

We can summarise the benefit which the shopper gains by retailers grouping in clusters by returning to the idea of the shopping list which needs to be ticked off as reliably and painlessly as possible. Shoe shops are the example par excellence of shops benefiting from locating alongside their competition and for the same reason as the ice-cream man. It is unusual to find isolated shoe shops. A town or shopping centre tends to support either several or none at all. When clustered together, shoe shops reduce the likelihood that the shopper will be disappointed and have a wasted trip because together they offer a greater choice than they could do on their own. That is a real benefit to the convenience of the customer. If there are advantages for shoe shops to locate together, there are even greater advantages in their being together with shops in complementary trades such as clothing and jewellery. Shoppers gain big advantages in being certain to get everything needed on a single trip. The journey to a regional shopping centre and back can take half a day and often more. To fail to find everything wanted is a significant loss of the shopper's precious time and money and one which it is worth making an effort to avoid.

Suppose a woman wanted to buy a pair of shoes, a blouse and a glass ornament as a gift and suppose she had the choice of two centres. The first was half-an-hour's travel distant and offered three shoe shops (including the shoe departments in big stores), ten clothing shops and a glass and china shop. The second centre was an hour away and also offered three shoe shops, but had more clothing shops and three large glass and china shops. Survey evidence suggests that increasingly shoppers would choose to travel further to the second centre because the chance of finding a suitable blouse and glass ornament would be greater. In this case the shoe shops in the second centre would benefit because of the greater choice offered by the non-shoe shops nearby. They would also be prepared to pay a higher rent for the privilege. The valuable synergy described here only applies between certain types of goods. Typically it does not exist with lower order goods such as food which are commonly regarded as a convenience chore, nor with branded or specialist hobby-type goods whose shops provide a destination in their own right. Nor does this story suggest that continued use of the town centre is essential. A large regional out-of-town centre would substitute for a town centre perfectly well.

Figure 7.2 shows that metropolitan centres have performed poorly. These are the top ten or so centres in the country such as Birmingham, Glasgow, Leeds and Newcastle. This suggests that there might be a size ceiling above which there may be only limited benefit for the shopper. In the last chapter it was pointed out that for out-of-town regional centres there was also in practice a ceiling above which few were built. From the shoppers' point of view there is a practical, physical limit to the area over which it is comfortable to walk. This can be partly off-set by big stores trading on more than one level and it is noticeable how first and upper floor trading is absent in smaller centres but increases rapidly once centres reach a certain level. Even with upper floor trading, however, there is a limit to the area over which a shopper is prepared to devote attention. Oxford Street, for example, is longer than most shoppers would want to walk and as a result there are many shops with more than one branch there, notably Marks & Spencer and Boots. In fact half of all shops belong to traders operating more than one fascia in the street.

From the retailers' point of view there is also a natural ceiling to size. As the size of centre rises, so too does the size of the catchment area, but beyond a certain point the additional trade attracted by increases in size starts to slow. Distance is a powerful deterrent and if the centre is so big that it can barely be covered by the shopper, little is gained by extending it further. Also if the shopper is given enough choice to be able to tick off all the items on the list satisfactorily, there is nothing to be gained from offering choice beyond that. The law of diminishing returns begins to apply. Oxford Street had 27 specialist shoe shops when surveyed in 1997, apart from the shoe departments of the big stores there. It is easy to see why a retailer might doubt whether a 28th shoe shop would attract sufficient additional sales to justify opening.

Finally it is worth ending this section on how clustering benefits the shopper with a word on branding. Branding generally favours dispersal and it is covered in the next chapter, but there is one aspect of branding which works in favour of the town centre and that is the value of a retailer's own name as a brand. In Britain, more than other countries, retailing is dominated by a group of national multiples which are household names. Their brand is often regarded by the shopper as more powerful than many manufacturer's brands. They sell products with their own label on them, but their brand is also felt to cover all the goods they sell whether they are own brand or not. Companies such as J. Sainsbury, Marks & Spencer, Boots and John Lewis enjoy a respect from their customers which leads them to believe that any item on their shelves, whether own label or not, must have passed a test of quality and value for money for it to be offered for sale. In locational terms, this strength of the retailer's own brand goes some way to off-set the move out-of-town which is so apparent among branded goods like cars and electricals.

The big high street names have invested heavily over the years in town centre property and though they have also opened out-of-town, the great majority of their stores are still in the town centre. Marks & Spencer, for example, most of whose stores are owned freehold, still have over 90 per cent of their stores in town centres despite devoting much of their new openings to out-of-town sites in the last 15 years of the recent century. This is a considerable investment to protect, and as long as retailers' brands retain their power it will continue to provide a powerful underpinning of town centre shopping.

Tenant mix

It is worth pausing at this point and having a brief look at tenant mix. The phrase 'tenant mix' refers to the policy of shopping centre owners in choosing and locating shops which come into their centres. There are two distinct parts to this, the choosing and the locating. The developer of a new shopping centre has the opportunity, usually with advice from specialist retail estate agents, to determine which tenants are or are not allowed into the centre. In having this power the developer has the advantage over the town centre where the mix of shops has evolved according to market forces limited only by planning and sometimes by terms which some landlords include in leases. Because a new shopping centre is let all at once, preferably before it opens, the developer has the opportunity to determine the composition of the centre and the mix of retailers included.

Most developers start with the anchors. These are usually large stores, department and variety stores in the case of a regional shopping centre. Once terms have been agreed with the anchors, the letting of the smaller shops can proceed. The shops are attracted by the promise of the anchors because they are not often willing to trade on their own. As a result the anchors exploit their bargaining power and the developer makes little or no profit from them, sometimes even a loss. In contrast the shops which come in after the magnets have signed up have to pay a high rent for the privilege because it is from them that the developer makes most profit.

Whether magnet or anchor stores are worth the privilege they are accorded is debatable. Department stores have tended to decline in recent decades and at the turn of the century all the big variety store chains, Marks & Spencer, Bhs, Littlewoods and C&A ran into difficulties. As we saw in chapter 6 not all types of shopping centre need anchors. The convention that the anchor stores provide the attraction to act as a destination is well established, however, and shopper surveys show them widely mentioned. Certainly it is reasonable to start with the largest stores when choosing tenants and then use these as an argument to persuade smaller and maybe less well known retailers to apply.

Tenant mix is not a scientific activity. Researchers have made attempts to answer questions such as what might be the optimum number of shoe shops to include in a centre, but such questions are difficult to answer with any precision or confidence. Shopping centres are not laboratories and control experiments are difficult at the very least. What tends to happen is that, with advice, the developer produces a marketing plan based on the size of centre, the quality of catchment area and the market position of the competition. This gives an indication of the balance wanted between different trades and whether the retailers sought after are up or down market. The appointed agents then try to implement the plan, but inevitably adjustments are often made during the course of letting. The developer can only choose from retailers who want to come in and are prepared to pay the required rent. In practice many a plan ends up changing because the retailers who want to come in may not be those who are wanted. It is a brave developer who leaves a shop empty rather than accept an offer simply because it fails to fit into the plan. Sometimes, however, rent may be sacrificed as when the offer from yet another shoe or fashion shop is declined in favour of a lower offer from a

Table 7.1 The shops of Bluewater shopping centre

	Number of shops			
	Upper level	*Lower level*	*Total*[2]	%
Larger Stores[1]	5	6	6	3
Clothing and Footwear				
Fashion general	18	16	34	
Fashion ladieswear	23	27	50	
Fashion menswear	5	11	16	
Childrenswear	7	5	12	
Footwear	6	8	14	
	59	67	126	36
Other Retail				
Accessories and luggage	7	12	19	
Books, cards and stationery	5	5	10	
Electrical goods	8	5	13	
Gifts	12	8	20	
Health and Beauty	2	7	9	
Homeware	8	8	16	
Jewellery	5	8	13	
Music and leisure	4	2	6	
Speciality food	2	0	2	
Speciality stores	0	4	4	
Sports and outdoor	4	7	11	
Toys and games	5	2	7	
	62	68	130	37
Services				
Cinemas	2	0	2	
Confectionary and ice-cream	4	8	12	
Dining	14	26	40	
Guest services	14	7	21	
Cash machines	6	4	10	
	40	45	85	24
Total	**166**	**186**	**347**	**100**

Source: *Your Guide to Bluewater*, March 2000

Notes
1 John Lewis, House of Fraser, Marks & Spencer, C&A, all on two levels with WH Smith on one level.
2 Three shops have more than one branch and five trade on both levels.

book shop. When this happens it is hoped that the benefit the centre gains from the extra range more than compensates for the short-term loss.

The second part of tenant mix refers to the placing of different shops around the shopping centre. The developer can decide where to encourage individual shoe retailers to locate, whether for example to cause fashion shops to group together or to disperse through the mall. Again there is little scientific evidence to guide the plan. There have been studies which have confirmed the not unexpected conclusion that fashion shops benefit by avoiding being next to markedly different trades such as hardware or butchers, but little that has helped in the main debate, which is whether a category such as fashion should be concentrated or dispersed.

In America there are shopping centres which call themselves fashion malls and concentrate on clothing at the expense, say, of furnishings, but they are likely to include fringe areas such as accessories and jewellery. In general the preference of most shopping centre owners is to achieve a good mix of trades and to disperse them evenly throughout the centre. The reason for this is the desire to achieve even pedestrian flow throughout the floor area. The design of the centre has this as a clear objective. This is why the anchor stores are placed at the edge of schemes rather than in the middle. The idea is that if an even plateau of attractiveness can be created the totality functions better than if some areas are perceived to be stronger than others. In practice this is extremely difficult to achieve and differences in pedestrian flow do occur which in turn lead to rental variations. Where weaker areas are inevitable they tend to be occupied by banks, building societies and other services.

Table 7.1 shows a simple analysis of the shops and services in the Bluewater shopping centre. Bluewater is the last and most sophisticated of the group of regional out-of-town shopping centres and opened in 1999. The table shows that there is a good mix between clothing, other retail and services. Compared with earlier centres it is noticeable that there are no less than 40 different dining facilities on offer. It is noticeable too that the different trades tend to be dispersed rather than concentrated. The table shows that in most cases the trades are divided fairly evenly between the upper and lower levels of the shopping centre. Bluewater is a good example of current thinking on tenant mix in practice.

Retail economics and shop rents

The benefits of clustering work to the advantage of both shopper and retailer. Retailers are able to buy into a position which gives them access to a concentration of shoppers walking by outside along the pavement or the shopping centre mall. In doing so they benefit the shopper by adding to the choice of goods available but they also benefit themselves. The only price they have to pay is rent. The payment of high shop rent enables the retailer to move from a secondary position to a prime location, and the way retail economics works it can often pay the retailer to make this move to compete for the best position. There is a form of leverage which tends to work in favour of high rents and helps explain why the slope of retail rental values in major town centres is so extraordinarily steep around the best position, as was shown in chapter 4.

The situation is explained in its simplest form in Table 7.2. Imagine a multiple retailer considering opening a new branch in a substantial regional

Table 7.2 The retail economics of two alternative sites

	Site A	Site B	
Sales	100	300	×3
Gross receipts (with 35% gross margin)	35	105	×3
Costs Rent	8	40	×5
Wages and other	20	40	×2
	28	80	×2.86
Profit (gross receipts less costs)	7	25	×3.6

centre. Two sites are on offer, A and B. B is the better located site but is otherwise identical to A in terms of access to parking, the size and shape of the premises and all the other factors which influence sales. The retailer has undertaken research of the type described in chapter 5 and concluded that B would generate three times the level of sales of A. Although this seems an enormous difference for two properties of identical size and specification, the difference is no greater than the retailer has come to expect from experience.

What they do find shocking is that the landlord is asking five times the rent for B when compared with A, even though B will produce only three times the level of sales. When they look at the profitability of the two options, however, they notice that despite this B produces the bigger profit. The gearing has had the effect of B producing a bigger increase in profit (3.6 times in the table) than the increase in sales. The reason this has happened is because costs other than rent do not rise pro rata with sales. The extra sales in site B will generate extra overhead costs but at a slower rate. The staff to service sales of 300 will be double, not treble, the 100 sales of site A. There are economies of scale at work here which mean that a shop in a prime location will work more efficiently than a less efficient shop with lower sales. The staff will be standing around doing nothing in A when in B they will be busier serving customers.

The same argument applies more strongly to other costs such as lighting and heating. In these cases there will be very little difference between the costs of A and B. The overall effect is that because costs other than rent rise less than sales between A and B, there is more room for rent to rise. Rent is best seen as a residual, as we saw in chapter 4. In fact the retailer could afford to pay slightly more than 40 in rent and still make a good profit. The gearing effect of costs rising more slowly than sales leaves room for the residual available to pay rent to rise more rapidly. This is what happens in practice and leads to fierce competition and often to redevelopment of valuable sites to increase the amount of attractive floor area. This increase can be achieved by bringing back land into retail use as often happens with comprehensive shopping schemes in town centres, or by rebuilding with upper retail floors as happens with department stores.

It all comes about because of the desire to cluster. Although weaker in some cases than it was, the desire to cluster is still extremely powerful. It leads to pressure for town centre redevelopment and to extremes of rental value. This pressure comes from the benefit it gives to shopper and retailer alike.

8   The desire to disperse

Minimising the trip

In describing the reasons for clustering in the last chapter it was argued that the journey undertaken to go shopping involves making an investment in time and money, and that having made that investment the shopper did not want to waste it by visiting a retail concentration which offered insufficient choice. If, however, the items on the shopper's list are regarded as routine and do not need time spent in choosing them as they did in our example of the wedding present, then the danger of disappointment becomes less. In this case the main concern is to reduce the effort of the journey. When the shopper knows precisely what to buy together with the price and specification that go with it, then shopping is simply a matter of acquiring it with the minimum effort. One way to do this is to stay at home and use mail order, direct mail or the Internet, but if a physical journey is involved then the objective is to minimise the effort of the trip, and this can involve preferring a dispersed location to the town centre. Traditionally the town centre offers both choice and convenience because there all the items on the shopper's list were near each other. Now, however, car ownership has freed the shopper to visit dispersed locations which are more accessible than the town centre with easier parking and lower prices.

The desire to disperse is the desire to minimise the effort of shopping, in particular the time and cost of the journey. The degree to which this force operates depends on the type of goods and whether the shopper regards them as routine convenience items or comparison goods. The economics of dispersal is driven by the cost of land and the associated costs of the retailer. There is no doubt, however, that the minimisation of effort is the fundamental force. One of the notable features of the decentralisation that has occurred during the last 20 years has been the size of the individual shop units built. Mostly it has consisted of supermarkets and retail warehouses, sometimes grouped into retail parks and shopping centres, but often free-standing. The minimum size is about 15,000–20,000 ft² (1,500–2,000 m²), but in practice little is built under 25,000 ft² (2,500 m²).

At this point, it is worth saying a word about the measurement of floor space. The floor space figure usually refers to the total area of a shop, called the gross leasable area. It is the measured area for which the retailer pays rent. The retailer will use part of the floor space for storage, offices and facilities for staff and use the remainder as sales area. The sales area is the part of the premises to which the public has access and it amounts usually to between 65 per cent and 90 per cent of the total. As distribution systems have improved this percentage has steadily risen at the expense of storage. Rather confusingly, supermarket floor space tends to be quoted in terms of sales area while other retail floor space is quoted gross. Because modern supermarkets include substantial in-store food preparation areas like bakeries, the butchering of meat and delicatessens, sales areas may amount to only 65 per cent of gross floor areas. Thus a supermarket with a sales area of 25,000–30,000 ft² (2,500–3,000 m²) could occupy 35,000–45,000 ft² (3,500–4,500 m²) gross.

Dispersal has produced quite substantial properties of the size which equal the retail floor space of a neighbourhood centre. By doing this the retailers have avoided one of the problems which had previously deterred dispersal. They place under one roof a range of functions which had previously been sold in different shops. Thus the supermarket replaces the specialist grocer, greengrocer, butcher and fishmonger. Large superstores are called 'one stop' shops because they offer facilities which enable the weekly household shopping trip to be completed at one place and paid for at one checkout. In addition to covering all types of food, larger superstores also offer standard clothing and household goods, cash points, post office facilities and sometime also a chemist, newsagent and café. Superstores offering this wide range of goods and services enable the shopper to tick off most of the items on the weekly list and do so under one roof. As a result they can have a devastating impact on small towns and suburban centres which had previously offered just such services. Big centres are less vulnerable because they also offer comparison shopping with which the superstore does not compete.

Free-standing retail warehouses are smaller than superstores but are still substantial buildings. Both DIY and retail warehouses selling electrical, furniture and carpets are able to offer space on a scale which the town centre finds it difficult to match. As with supermarkets, however, retail warehouses were able to counterbalance the loss of choice which shoppers may have suffered by leaving the town centre by providing bigger choice themselves in one store. Carpets, furniture and electricals are all durables where comparison shopping takes place. Until 20 years ago, this form of retailing took place in town centres forming part of the traditional hierarchy. When they moved from town centres to out-of-town they mostly did so to groups of similar firms on retail parks. DIY, in contrast, had no preference for either centres or free-standing locations. The main advantage for the retailer was in having a much larger floor space than was available to them in the town centres. The increase in space was by a factor of five to ten. They were therefore able to supply a much greater range of goods than before. From the shopper's point of view all that was lost was the ability to combine furniture or electrical goods shopping with other types of shopping. But since these are rarely purchased items this was a price worth paying in return for the greater choice which came from the huge increase in floor area.

Out-of-town locations usually have main road frontages. A typical advertisement by a retailer looking for a free-standing retail warehouse site would ask for a flat, rectangular building with a main road frontage covering half of a site of 2–4 acres to allow for ample car parking. Such sites are usually found on roads radiating out from town centres which allow the retailer to intercept trade going into the town centre. The ability to intercept flows of trade particularly if free parking is provided is one of the characteristics of dispersed retailing.

Interception occurs where the retailer is able to take a position so as to offer better access to the shopper travelling to another destination. This is often found in holiday areas of France where the local authority has provided lay-bys where traders are invited to sell local produce on roads leading to popular beaches. It is not uncommon to find other traders positioned in unofficial roadside sites lying ahead of the purpose-built lay-bys (see Figure 8.1). In doing so they not only avoid paying a rent, they also attract the attention of the holiday-maker ahead of the official site. In the same way out-of-town retailers can intercept the flow of shoppers going to the town centre and can do so with lower costs than they could in the town centre.

Interception has become more relevant as population itself has decentralised and people abandon the main built-up area of a town to live in the outer suburbs or surrounding villages. This population movement has meant that the average distance from house to town centre has tended to increase making the shopping trip to the centre more vulnerable to interception. In fact edge of town retail development is often nearer to areas of new housing than to the town centre.

The effect of branding

At first sight it might appear that branding has little relevance to the location of shops, but in fact the effect is far from negligible, and branding tends to work against comparison shopping and to encourage decentralisation. Branding is the process whereby a line of goods is made uniform and given a name. It means that a tin of Heinz baked beans or a Mars bar will be identical and interchangeable with any other. For the manufacturer branding has the advantage of longer production runs and being able to build customer loyalty. Brands can be valuable in themselves and there is a whole branch of marketing literature devoted to their study.

From the shopper's point of view, a brand offers reassurance, the promise

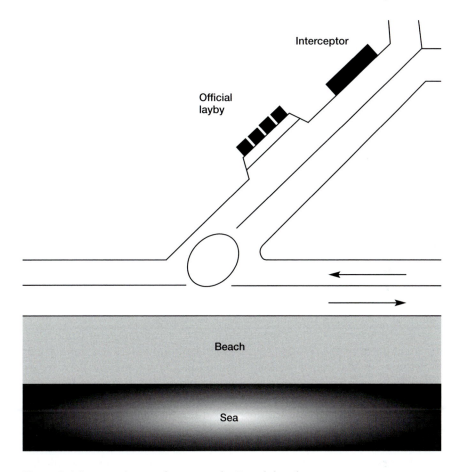

Figure 8.1 Interception on the way to the French beach

of not being disappointed. Once a branded product has been tried and found satisfactory it can be bought again without any feeling of anxiety as to its quality or what it consists of. It reduces the searching process for the shopper and this is a real advantage. As a result of this, branding often has the effect of causing the precise purchasing decision to be made before the shopping trip rather than during it, and it is why branded goods are so heavily advertised on TV and in magazines away from the point of sale. In other words the shopping list has a branded name on it rather than a commodity. Branding is thus the enemy of comparison shopping where a shopper uses skill and taste in selecting goods. When it is a case of choosing a cocktail dress, a new tie or even selecting fruit of the right degree of ripeness, there is no substitute for most people in being physically present and making a comparison and a decision on the spot. Branding is of little help in this type of shopping. As branding grows, as it has in the 1990s, particularly in clothing with the expansion of American style designer labels, the less becomes the role of comparison shopping and the need for clustering.

The effect of branding on retail location was first felt with more expensive items such as cars followed by electrical goods such as TV and hi-fi. Here the purchase is of sufficient importance that the brand is chosen after considerable time and thought, but this process takes place before visiting the point of sale. The selection process may involve discussion with friends or seeking advice from specialist magazines and similar sorts of information. Once the decision has been made, however, the shopping trip is reduced to comparing price, and more secondary factors such as delivery time, guarantees and servicing. This means that the buying decision has been made before the journey to buy takes place and the advantage to the shopper in having the relevant shops clustered together is considerably reduced. As a result cars and electricals (in particular the more expensive white goods) have abandoned city centres in favour of retail warehouses outside. There are some examples of American style automotive strips where car showrooms gather in loose groupings, but on the whole cars and electricals today retail from free-standing locations.

In fact the car was the first type of retailing to decentralise, moving out of the high street in the 1950s and 1960s. There are still a few luxury car showrooms in Mayfair, in Berkeley Square and Park Lane but they are so different from the modern car showroom out on the ring road that it is difficult to recall that most car retailing was once in that form. The reason cars pioneered the move out-of-town was because they needed more space than they could find in the town centre and because their customers by definition had (or were about to have) cars and so had access to them. Once price becomes the overriding factor then the retailer takes steps to minimise that price by choosing a site where land is cheap and space sufficient to allow economies of scale, which in practice means a move away from the town centre to out-of-town. Department stores have complained that shoppers will visit them to make use of their knowledgeable sales staff and to take away manufacturer's catalogues and then go off and buy at a discount store which offers no such services but a cheaper price. It is not surprising therefore that electrical retailing, another area of branded goods, followed cars as one of the earliest forms of retailing to desert the town centre. Little remains today of electrical chains such as Rumbelows, Currys and Electricity Board showrooms which were once prominent in the high street, although Currys now operate out-of-town as part of Dixons.

Branded goods range from small grocery items (Swan Vestas matches) through to Rolls Royce cars, but the dominance of branding varies widely from areas where it is universal (electricals) to areas where it is weak such as fruit and vegetables. It is weak where the uniformity of the brand offers little value, such as fresh food, jewellery, and antiques and household ornaments. One area where branding has traditionally been weak is clothing. The huge variety of styles, not to mention size and colour makes the long manufacturing runs associated with branding harder to achieve. The result has been for retailers themselves to fill this gap by offering the comfort of their own brand. In fashion, names such as Top Shop, Jaeger and Laura Ashley have identified themselves with segments of the market so that the shopper knows the type of clothes on display before she enters the shop.

It is worth noting that Britain is unusual in this respect. Fashion in other countries is not dominated by national chains to anything like the same degree. In continental Europe the role is filled by independent boutiques, while in America the designer label is important. In both places chains of fashion shops are weaker than in Britain.

In the 1990s there was a big expansion of designer fashion houses. For the first time a number of names developed into world-wide brands functioning on a global scale. Led by the American houses of Ralph Lauren, Donna Karan, Calvin Klein and Tommy Hilfiger, and by a group of Italians, principally Armani and Versace, these firms have invested heavily in building a world-wide network of outlets. They began in haute couture at the very top end of the market, then extended to ready-to-wear and diffusion. Ready-to-wear involves marketing the haute couture designs to a wider market, but it is the move into defusion which really exploits the value of the brand. Diffusion involves producing low value merchandise such as jeans, T-shirts and accessories and allows the designer to enter the mass market without diluting the exclusivity of their haute couture business. Diffusion goods are sold at a premium and probably amount to the most extreme value placed on a brand found anywhere in retailing.

Calvin Klein jeans and Tommy Hilfiger T-shirts, however, are branded goods like any other and as a result this encourages retailers to sell them wherever costs are lowest. This amounts to a trend in favour of dispersed locations because the advantages of choice offered by clustering have been reduced.

The new strength of brands in clothing is already apparent in factory outlet centres (see chapter 6). The whole point of these centres is to sell branded goods, to offer the shopper the chance to buy a valued brand at a discount. Without branding they couldn't exist and they are all located outside traditional centres. They have to be away from town centres where the brand is on offer at full price so that it isn't undercut.

Designer fashion is sold through flagship stores which are mainly limited to upmarket locations such as Bond Street and Sloane Street in London and Fifth Avenue and the Upper East Side in New York. Ready-to-wear is sold under franchise in many department stores and the defusion lines of jeans and T-shirts more widely still. With the strength of the British fashion multiples there may be some doubt as to whether the British buyer is prepared to pay a premium for a designer label, particularly on more basic clothing items, but the marketing effort of the big designer houses seems likely to push clothing in the direction of other branded goods. The result of branding is to emphasis price and convenience at the expense of comparison and clustering. It tends to move retailers away from town centres towards the lower cost, more dispersed form of retailing found in America. After all America is where a very large number of international brands originated in the first place.

The three waves of decentralisation

In this chapter we have examined the causes of retail decentralisation. It complements chapter 2 which considers the political debate which took place. The rise in car ownership, the movement of population further from town centres and the effect of branding are all long-term trends which have been at work for decades to reinforce the natural desire to disperse which is ever present. In Britain these trends combined to cause a dramatic change to the retail landscape which occurred in the last quarter of the twentieth century. This change can be described in terms of three waves of decentralisation. The first wave involved food. During the 1970s supermarkets began opening larger new stores outside town centres, often paid for by the sale of their old town centre shops. Food retailing was undergoing a revolution with the introduction of self-service to replace counter service. This in turn led to the use of a wheeled trolley to replace the wired basket. The combination of increasing car ownership plus self-service and the use of refrigerators made it possible to shop less often but in bigger quantities.

The effect of this was for goods to be sold in bigger pack sizes. New types of product appeared such as kitchen rolls and fruit juice which were far bulkier than had been possible before. The shopper who could wheel the trolley direct to the boot of the car had a huge advantage over the shopper who used the bus or walked. In particular women accompanied by young children were shown to be vulnerable. Unless there was a clear single level from checkout to car park (often not the case in the town centre), they were faced with trying to negotiate stairs and non-automatic doors with a pushchair and a load of heavy shopping. The advantages in the shopper being able to wheel a trolley direct from the checkout to the boot of the car was so obvious and overwhelming that they proved difficult to resist. Indeed the movement out-of-town by food was so rapid that it often left the town centre denuded of food shops. The 1990s have seen supermarkets returning with smaller specialist stores to town centres. In Oxford Street, opposite John Lewis, is a Tesco Metro store. For most of the 1980s and 1990s that site was occupied by members of the Burton clothing group but before that there was a Woolworths there which devoted a floor to food sales. Thus in 20 years this prime site has move from food to clothing and back again to food. By the mid-1990s the building of new supermarkets began to slow and there was talk of saturation. Smaller stores began to be opened near small market towns which previously had offered an inadequate market. The government commissioned a consultant's report which was published in 1998 (CB Hillier Parker and Savell Bird Axon 1998). This showed considerable levels of impact:

> Our research shows that the impact of large out-of-centre and edge-of-centre food stores is not limited to convenience retailing, but can also adversely effect comparison and food uses. For example, in Fakenham the reduction in turnover of six comparison retailers ranged from 3.7 per cent to 18.9 per cent.

The second wave involved retail warehouses. It arrived five to ten years after the first wave and reached its height in the late 1980s. It included bulky goods such as furniture and carpets, and larger branded goods such as white electricals. These categories were joined by DIY, a largely new form of retailing. Retail warehouses have an industrial background and first began to appear on a large scale in the early 1980s when there was a major contraction in manufacturing. Retail was one of the few alternative uses possible for the large amount of newly vacant industrial property. There followed an enormous increase over a short period of time in retail warehouse and retail park openings. This may have had less impact on public perception than the rapid spread of out-of-town superstores, but from 1985 to the end of the century retail park openings were the dominant form of retail development. In 1985 retail park floor space past 1 million ft^2 (100,000 m^2). By 1990 the figure reached 26 million ft^2 (2,500,000 m^2). To put that in context, the total floor space in all other types of shopping centre which opened between 1985 and 1990 was 27 million ft^2 (2.7 million m^2). During the 1990s while superstore and other types of shopping centre openings slowed, retail park openings continued to dominate. Since 1993 over half of all new shopping centre openings have been in retail parks. No longer are retail parks dominated solely by DIY, furniture, carpets and electricals. A whole raft of other types of retailing from computers to pets have made them their home.

The larger town centres rode the second wave of decentralisation as comfortably as they did the first. Although electricals, furniture and carpets were clearly comparison goods, the heart of comparison shopping was clothing and this was still untouched. Oxford Street, for example, in 1983 contained 226 retail shops of which 153 sold clothing and footwear. It was a considerable shock, therefore, when in 1984 the nation's largest clothing retailer, Marks & Spencer, announced that it was proposing to open out-of-town stores. The third wave of decentralisation involved clothing, footwear and the type of comparison goods commonly found in town centres. Unlike the other two waves it is possible to date its start in a precise moment in time, the AGM of Marks & Spencer in May 1984. In their annual report the company complained of local authorities failing to accommodate car-borne shoppers and announced its intention to develop out-of-town stores. The three waves therefore occurred within a decade and in that short period of time the attitude to out-of-town by retailers, developers and investors, not to mention planners, was transformed.

The coming of the third wave of decentralisation in 1984 was followed by a more laissez-faire attitude to central government which allowed market forces to have their way to a greater extent than before or since (see chapter 2). Retailers and developers responded to this favourable environment by bringing forward proposals for out-of-town development. Although proposals for regional centres were largely stopped, there was an explosion of new retail parks which included within them many comparison retailers who were hedging their bets.

Retailers, who a few years before had been content with a town centre location, now began actively to plan out-of-town formats. The problem they faced was that there were very few out-of-town regional centres available. In fact only four opened in this period, and the alternative, the retail park, was unfamiliar and at first sight appeared unattractive. Also, for retailers used to trading in unit shops, the retail warehouse was often simply too big. Nevertheless there is no doubt that the town centre, in retailers' eyes had lost its monopoly as the only suitable location for comparison shopping.

The town centre has survived this loss of monopoly well. Mainly this has been due to a remarkable expansion in the 1990s of eating out, with many large prime, well-located high street properties such as redundant banks being converted into wine bars and themed pubs. Though out-of-town has also benefited from this trend, the town centre has taken the lion's share by offering available property and an attractive volume of customers.

The 1990s saw a change to a more hostile attitude to out-of-town development but it has proved to be difficult to stop. Retail parks, in particular, continued to be the dominant form of out-of-town development. The result has been that though market forces might prefer regional centres, in practice the retail park or a free-standing site on derelict industrial land was often the only alternative. This has led to comparison retailing occupying sites on suitably designated retail parks such as Fosse Park near Leicester. Marks & Spencer, for example, in their catalogue of October 1999 list 18 out-of-town stores among 44 named as being 'our largest stores . . .'. Of these, 18 seven are in regional centres and 11 are free-standing or in small centres. This somewhat confused picture of partial decentralisation and a split between centres and dispersed locations would be found among many of the country's leading comparison retailers.

The position at the end of the century after 25 years of rapid change is that the town centre survives although weakened. Decentralisation continues, though at a more modest pace than before. Those town centres which have done best are those which have encouraged large-scale retail development and which also offer an attractive environment. The full impact of the three waves of decentralisation over 25 years is still making itself felt and will lead to the contraction of many town centres. Those which do best will be those which offer the shopper more than simply the basics of a big cluster of shops but also offer alternative leisure facilities which go to make the total experience of shopping a pleasant one. It is this aspect of shopping which is the subject of the next section.

Retailing and leisure

Up until now shopping has been described as a chore to be minimised. This is true of most shopping but by no means all. Many people enjoy browsing in bookshops, poking about in flea markets or looking at art and antiques. In addition, for most people there is an area of special interest and expertise which gives pleasure. This might be a hobby but could also be a wider area of interest than the word hobby implies. Within that area shopping is often an enjoyable activity and people's attitude towards it is the opposite of a chore. We will call it hobby shopping for short.

In addition to hobby shopping there is another type which also gives pleasure and is the opposite of a chore. This is leisure shopping which occurs as part of a trip for some other purpose. It might involve a day out to attend an event, visit a site of historic or picturesque appeal or a walk in the country. Such a trip might include a stop at a small attractive town for lunch followed by a browse through the town centre. This type of trip is becoming steadily more popular with widening car use and the leisure to indulge it. From the shopping perspective the point to note is that the trip functions as the reverse of the trip described earlier in this chapter. Far from seeking to minimise the journey, the leisure trip is itself part of the appeal, along with the freedom to go and stop wherever wanted. Also shopping is not the prime purpose of the trip but an incidental, a casual occurrence of impulse purchases. It is shopping without a list.

It is difficult to measure, but it is probable that spending on hobby and leisure shopping is growing at a much faster rate than retail sales in general. Certainly the statistics show expenditure on discretionary items and items 'not elsewhere classified' to be growing faster than routine items. The enormous success of Covent Garden and its many imitators supports this. From the point of view of retail location, the importance of this growing part of retailing is that it is as fancy free as the shoppers who take part in it. It is divorced from the hierarchy, its location not so much out-of-town as independent of the whole town centre versus out-of-town dynamic. Many of the places where this type of shopping thrive are historic town centres, the waterfront of converted dock areas, or buildings converted from another use such as Covent Garden or Albert Dock, Liverpool.

One can see the power of this type of shopping from an examination of shop rents and the presence of multiples. Since the success of the festival mall at Covent Garden, the neighbouring streets have attracted more conventional shops willing to pay rents so high they match parts of Oxford Street. Streets such as Neal Street have changed in 15 years from being minor

Table 8.1 Heritage town ranking: population compared with retail size

	Population		Retail size	
	Number	*Rank*	*MC* [1]	*Rank*
Oxford	133,000	52	104	12
York	127,700	56	97	18
Cambridge	122,600	60	104[2]	12
Chester	89,900	90	100	16
Bath	87,100	93	89	27

Sources:
Population: 1991 Census *Key Statistics for Urban and Rural Areas*, Table 1: Individual urban areas, population present
Retail size: Hillier Parker (1995)
Notes
1   MC = multiple count, i.e. the number of national multiples present in the centre.
2   Cambridge figure may be inflated because the centre is split between Petty Cury and the Grafton Centre, leading to several multiples having branches in each.

in retail terms to being among the highest rented in Britain. As Table 8.1 shows the five heritage towns of Bath, Chester, York, Oxford and Cambridge all are ranked far higher as shopping centres expressed in terms of national multiples than they are in terms of population.

Similarly many large towns have less luxury shopping than do neighbouring small towns in high income areas. Thus Slough has only a fifth the number of luxury facilities compared with Beaconsfield. Similar relationships can be seen in, for example, Beverley when compared with Hull and Harrogate and Leeds.

Finally in this section it is necessary to mention the growing importance of eating out. Eating out and take-aways have grown so greatly that in America the role of food and supermarkets in the retail scene is far less than in Europe. The cause is clearly the great increase in the number of married women with full-time jobs and the decline in the formal eating together of families which has accompanied it. Europe is rapidly following the American path and this can be seen in the huge increase in the proportion of all types of shop property which is occupied by catering establishments of one type or another. Pubs, cafes and restaurants are beginning to cluster together and establish desirable 'pitches', following in the pattern of shops. For many town centres their expansion has compensated for the decline in conventional retailing. In fact their location behaviour is similar to that of shops which serve the leisure shopping just described, that is, they prefer an attractive environment. An examination of the occupiers in many new waterfront developments shows that half or more may offer food or drink. Often it appears that it is the catering offer itself which is the magnet rather than the shops selling goods.

The three types of shopping dealt with in this section, hobby, leisure and eating out, are important because all are growing fast. Hobby shops are often content with free-standing or out of centre locations, but leisure and eating out are as likely as not to group together and it might seem strange that they are placed in a chapter dealing with dispersal. The reason for this is because any grouping that occurs is outside the normal grouping of shops. In this they resemble the other forms of dispersed retailing described in this chapter. The factor which determines the location of this type of shopping is not accessibility (called centrality in central place theory), as it is for clustered shopping centres, but rather an attraction such as a waterfront, a cathedral or a beautiful townscape. The location of this attractive factor has nothing to do with retail location and whether it lies within an existing town centre is incidental. That is why we must consider the growth of leisure shopping to be an example of dispersal rather than clustering.

In fact much of the growth of this type of shopping has taken place in town centres and has been invaluable in lessening the rate of decline which would otherwise have taken place. There are two main reasons for this. The first is that town centres are often attractive places to visit apart from the everyday shops they may contain. The second is that the decline in town centre retailing has made property available for this growth. As we saw above, eating out in particular can benefit from clustering which in practice usually takes place in town centres. For town planners this is the best opportunity they often have to revive a declining town centre.

9   Office location

Ways of work

Most people work in offices, if you draw the definition loosely enough. An office is a room designed for working in. It contains the basic walls, windows, doors and ceilings, and has the ability to accommodate such supporting equipment as can fit in a reasonably-sized space (say 5,000 ft^2 (500 m^2)), and is compatible with a person working at a desk. Such a definition fits a type of building usually called an office block and if it is accepted then it follows that the people working there are office workers whatever they may actually do. It includes activities which are not usually thought of as office work such as dental surgery or a college or TV station. In cases such as these there may be some uses which require a purpose-built specialist building, such as an auditorium for a college or a studio big enough to accommodate scenery for a TV station, but the majority of functions are suitable for a conventional office building.

This has become increasingly the case as equipment has become lighter, smaller and less demanding of special space. A good example of this is computers. When they first started to be used on a wide scale in business, computers meant massive mainframes. Their weight was so great that the floor loading of the average office block was inadequate to bear it and special reinforced buildings were needed. Further, the computer room had to be air-conditioned and kept at a constant temperature and was a forbidden sanctum guarded by men in white coats. In the 1980s this was no longer necessary but the VDUs still needed substantial ducting which required raised floors or false ceilings and problems with walls and columns. This made it difficult to install computer systems in offices which were in converted historic listed buildings, of which many are in Mayfair and other high rented parts of central London. By the late 1990s with the Internet and the laptop, the demand for special facilities fell again and the need to adapt buildings for computer use fell with it. People no longer think it odd that the computer fits into offices as easily as equipment such as a typewriter. Indeed typewriters have themselves been replaced by computers.

One of the results of this trend has been for more varied uses to be found in buildings that to outward appearance are conventional offices. To help this trend along new offices have increasingly been built to shell and core, that is the construction of interior walls and corridors has been left until the building has been let so that they can be finished to meet the requirements of the tenant. This has given greater flexibility to accommodate the needs of an ever widening base of tenants. It is a common fallacy to think of offices as paper factories full of clerks engaged in the process of administration. In fact this function has been reduced by computers but in its place has come an ever widening group of office users which cover almost any activity which can be thought of. Le Corbusier called a house 'a machine for living'. The office has become a machine for working.

Types of office user

In recognising the flexibility and ubiquity of offices, it follows that the behaviour and locational need of office users is wider than traditional analysis allows for. Up until about 1975 such minimal academic interest as there was in offices regarded them as supplying an administrative service to manufacturing industry. The baleful influence of manufacturing on locational theory continued long after manufacturing became a small minority of the economy. By the end of the recent century manufacturing accounted for barely a quarter of the capital value of firms listed on the stock exchange. This excludes the 40 per cent or so of GDP which is taken by the public sector. Manufacturing administration, therefore, accounts for only a quite small proportion of office occupiers.

Alan Evans, writing in 1985, speaks of three types of office users: finance, corporate head offices, and business services. These three uses are important because they tend to operate on a large scale. Although, as we shall see in this chapter, their locational needs are varied, they include that interesting group of users who seek prime central location and are prepared to pay very high rents for the privilege. The Americans refer to FIRE, standing for Finance, Insurance and Real Estate as the key group of central business district offices users.

Writing today it would be necessary to break down finance and business services to distinguish those which tend to centralise and those which disperse. It would, however, also be necessary to make additions to the list. The media has emerged as a major user of central office space. This includes not only newspapers and television but also publishing, literary agents and a whole range of ancillary activities. Newspapers used to be famed for their concentration round Fleet Street, but today scarcely any remain there. Because of its desirable location and the fact that newspapers occupied larger than average sites, redevelopment occurred for occupation by finance and business services. The newspapers themselves dispersed to Docklands (*Daily Telegraph*, *The Times*, *Daily Mirror* and the *Sun*), Southwark (*Financial Times* and the *Express*), Kensington (*Daily Mail* and *Evening*

Standard), and elsewhere. The desire by different activities to cluster within central London appears to be weaker than it once was and is not strong enough on most occasions to withstand development pressures, as happened with newspapers in Fleet Street.

Mention should also be made of the entertainment industry which concentrates heavily in a few major world cities. The recording studio is another example of an activity which because of lighter and more flexible hardware is now more likely to be found in a conventional office building than it was 25 years ago.

There are two other office users which should be identified as tending to locate in the centres of capital cities and they are central government and flagships. Despite great efforts to decentralise, central government is still heavily concentrated in central London, and the size of the government office estate has hardly changed in several decades. Flagship activities include the headquarters of a myriad range of institutions such as the offices of professional bodies, and other organisations ranging from the Church Commissioners at Millbank to the NSPCC (in the City). A central London flagship is useful for such bodies because it provides a good base from which to lobby and listen to government and the media and also provides an accessible location for their membership. There are several hundred of such non-commercial flagships occupying space in central London. For them the advantages of a central location are very strong, and being mostly small, the total cost and the high rent they pay is tolerable. We can therefore identify six types of office users who are found in high-rented central areas: finance, corporate headquarters, business services, media and entertainment, central government and flagships.

Why pay more in central London?

Using our wider definition of offices it is possible to identify a huge number of different users. All serve a customer base ranging from local to national. Thus a small town might contain a chiropodist, a solicitor and an estate agent, while a county town would in addition contain the head or regional offices of local businesses plus local government and regional offices for central government and its agencies. The reader may notice a similarity here with the description of the retail hierarchy in chapter 6. As with retail, however, in addition to the local-to-national hierarchy there is also the centralised–decentralised dynamic. Some offices locate in town centres, others disperse to the edge of the city. This applies mainly to central London, but the process can also be seen round office centres in the rest of the country. Since the introduction of the B1 use class in 1987 (see page 114) businesses have been free to locate either in a city centre or a dispersed business park. Thus offices in, say, Reading are divided between the town centre and a variety of business parks on the fringe and no one is forced to choose one in preference to the other. Office users are subject to the same push and pull as retail between the desire to cluster together and the desire to disperse. In the following section we look at why this is so and how different users are affected by these two fundamental forces. Let us start here with the six types which were found in central London.

There are a number of factors which apply to all six types of central London office user. All need to have the ability to pay the high rent and other costs. However strong the desire may be it will not be satisfied unless this ability exists. Inertia is also undoubtedly a factor. Property leases are commonly for 25 years, often longer. Once property is obtained the rent paid may be below the open market level. This combined with the cost and disruption caused by a move out of central London is undoubtedly a factor in favour of staying put. All six users benefit, too, by having good access to the rest of the country. In many organisations there is a need to hold meetings of people coming from all over the country, and London, because of its transport links, is in practice more accessible than any other part of Britain. Fourth, all users benefit by the proximity of a large number of important people in other fields. London is the centre of political power and controls the supply of capital. For the flagship users this was referred to as lobbying, but it is also listening to what is happening outside the organisation's immediate environment.

This involves communication with a whole range of different people. Despite the enormous progress made in electronic communication, most of this contact involves meeting and conversation. Central London could thus be seen as one gigantic talking shop. Studies of the behaviour of organisations shows that the higher up one goes the greater is the proportion of time that is spent talking rather than gazing at a screen or performing some other task. In the 1970s Goddard undertook a number of diary studies of senior executives to track the interaction they engaged in. This showed that only the most senior staff of a corporate headquarters could justify a central London location, and the support staff could decentralise. In fact this has largely happened since then, but the steadily increasing ratio of chiefs to indians in London has meant that the amount of talking taking place has continued unabated.

Evans, in the early 1980s, was able to show that a high proportion of the largest corporate head offices were to be found in London (70 per cent of the top 100 companies). The equivalent figure in 1998 is 64 per cent, but this is misleading because there has been a big move out of the centre and today these big companies manage either without a London head office at all or with just a small boardroom suite which enables the lobby-listening activity to take place (Table 9.1). Big companies are frequent visitors to central London to buy services, but they do not need a permanent physical presence to do that. The cost is too high.

Central London is the centre for both government and finance. In some countries these two functions are in different cities as for example in Germany (Berlin and Frankfurt), Italy (Rome and Milan), and America

Table 9.1 Examples of corporate head offices in Central London

1 *The dying breed of companies maintaining a traditional head office function**	
Shell	The Shell Centre, South Bank
ICI	Millbank
Unilever	Blackfriars
Marks & Spencer	Baker Street
Prudential	Holborn
2 *Companies with small head office suites*	
Cadbury Schweppes	Berkeley Square
GEC	Stanhope Gate, Park Lane
Glaxo Wellcome	Berkeley Square
Guinness	Portman Square
Allied Domecq	Portland Place
Rio Tinto	St James's Square
Bass	North Audley Street, Mayfair

Note
*Several of these companies have announced that they are actively considering moving.

(Washington, DC and New York). These two functions account for the presence of the media and a whole range of other organisations which depend on them and service them. They are markets and magnets for information, money and power and they operate on an international as well as a national level. Thus government is surrounded by foreign embassies while the Bank of England attracts several hundred foreign banks. These markets and magnets are so powerful that they cause the employment of several hundred thousand workers who are dependent on them and are in consequence tied to central London. Finance and business services together account for half of all employment in central London, with finance fairly static and business services growing fast. These workers and their firms in turn benefit from the scale of the operation within which they operate. At an individual level there are alternative jobs and the opportunity for promotion. For the firm there is similarly the opportunity to expand and develop. There is also a large pool of specialist labour from which to recruit staff. In short there are benefits to be had from the massing of employment which surrounds the two poles of power.

For readers unfamiliar with the City of London it is worth giving a word of explanation of the financial function. First, there is the Bank of England which lies at the traditional hub of the City and which functions as a regulator and arm of government, and also as the pinnacle of the banking pyramid. Surrounding it are markets for stocks and shares (the Stock Exchange), currencies (foreign exchange), specialist insurance (Lloyds), futures and other derivatives (Liffe etc.), shipping and a whole range of commodity markets covering metals, oil, tea and coffee. In the next ring lie specialist services such as investment (or merchant) banking which deals with take-overs, mergers and flotations, together with fund management.

These services operate internationally and generate much wealth and employment.

General insurance used to be added to this list, but insurance found it could function adequately outside central London and has been a heavy decentraliser in the last third of the century with Pearl, for example, moving from Holborn to Peterborough. Many of the other functions described might also decentralised, particularly now that many markets operate electronically with deals made on computer screens rather than on trading floors. Indeed the success of Docklands in attracting financial services away from the City suggests that the need to cluster is less strong than the need to enjoy modern offices with large floor plates. For over three centuries the City has functioned successfully as a huge anthill of interlocking markets and services. Financial services have expanded there so that other uses have gradually been excluded. Up to the Second World War docks and ware-housing, manufacturing and residential continued to occupy considerable space. Today the City is as much a mono-culture as any modern farm with few non-financial uses remaining. Yet the City has shown itself to be subject to the pressures to decentralise. It will be interesting to see how the future develops. Apart from the move to Canary Wharf and Docklands there has also been a move by stockbrokers to the West End. The American experience has been for Wall Street to decline in the face of general decentralisation and competition from Mid Town (the New York equivalent of the West End).

Finally, a word needs to be said about business services. These are services which are offered generally to business rather than being particularly focused on any particular business sector. Lawyers, the big accountants and chartered surveyors are examples, together with computer software, advertising, PR, market research and a whole range of specialist consultants. These firms themselves often grow to a large size and operate inter-nationally, and they have grown steadily over the recent decades. It is worth pointing out here that though business services are growing in central London, they and particularly financial services have grown even faster in the remainder of London and the South-East.

Unlike the other office users (see newspapers above) the major business services often still group together within a particular district of the capital. Thus lawyers concentrate around St Paul's and the Barbican between the Bank of England and the law courts and chartered surveyors are found in the north-east corner of Mayfair between Hanover Square and Berkeley Square. These concentrations are analogous to clustering by comparison retailing. They occur because it is to their advantage and to the advantage of their customers. Certainly where there is an element of a market existing as occurs in the case of investment property, where the market is held by the leading firms of chartered surveyors, it is easy to see why it is to the advantage of the customer to have the main market players close to each other.

In summary it is possible to identify seven factors which encourage office users to locate in central London (and by inference in the centre of other cities). Four of these are general factors which apply to all users. There is the ability to pay, and the inertia of those already there. In addition to this come the lobby and listen function and the fact that a central London location provides maximum accessibility to the remainder of the country. There are in addition three factors which do not apply universally. Government and finance act as markets and magnets, and several of the functions benefit from the massing of a pool of labour and a concentration of money and power. These would include finance, business services and the media. Finally, several types of business services, as we have just seen, by forming local clusters, act as markets serving a national and international clientele. As a result they tend to be tied to a central location.

The seed pod model

Despite the seven factors leading to a central London location which were summarised above and are set out in Figure 9.1, decentralisation has been occurring at a steady rate at least since the 1960s. There was a feeling then that too many offices were situated unnecessarily in the capital causing wastefully high costs, not least in long distance commuting. The George Brown ban on new office development in the South-East without a special office development permit was introduced in 1965. At that time, too, the Location of Offices Bureau (LOB) was set up to encourage decentralisation. Posters appeared in the London underground to show the delights of working in the countryside and pointing out the costs of a wastepaper bin in the City. LOB statistics (1963–80) showed the annual level of jobs moving out of central London. The series is now kept going by Jones Lang La Salle and they showed that in the recession (1990–92) 36,000 office jobs were relocated from central London. Although central government moved many agencies away from the South-East, for example the Driver and Vehicle Licensing Agency to Swansea, and the headquarters of the National Health Service to Leeds, it was found that private sector firms were reluctant to move so far. For large firms the move usually involved a split between the decentralising back office functions and a small head office which remained in central London. The need to communicate between the two offices was such that the back office rarely moved more than a 100 miles or an hour-and-a-half's rail journey from London.

Insurance companies formed a large proportion of the bigger decentralisers in the 1960s and after. By 1973 it was possible to identify 64 different moves by insurance companies to destinations within this ring. Bristol, for example, attracted Phoenix, Legal & General, Prudential, Eagle Star, London Indemnity and Sun Life. Not all these moves, it should be noted, were to town centres. Friends Provident occupy a campus site outside Dorking as do Pearl in Peterborough. Small companies, where it was

General factors	Special factors
• Afford to pay • Inertia • Lobby and listen • Access to country	• Benefits of massing • Markets and magnets • Local cluster

Central London office user types

• Finance • Government • Corporate HQs	• Business services • Media • Flagships

User types to which special factors apply

Special factors	User types		
Benefits of massing ⟶	Finance	Business services	Media
Markets and magnets ⟶	Finance		Government
Local cluster ⟶	Business services		

Corporate HQs
All plcs

Flagships
Professional bodies
Trade associations
Trades Unions
Special interest groups

Government
Ministries
Agencies
Quangos
Embassies

Other
Colleges
Hospitals
Medical research
Libraries

Finance
Banking
Stockbroking
Futures etc.
Foreign exchange
Insurance
Commodities
Shipping
Fund management
Pension funds
Other investors
Property investment
Property management

Media
Newspapers
Magazines
Book publishing
Television and Radio
Art and Antiques
Cinema and Theatre
Music

Business services
Law
Accountancy
Audit
Management consultancy
Pension management
Personnel
Computer software
Advertising
PR
Marketing
Market research
Design
Property agencies
Property valuation

Figure 9.1 Central London office users and factors which attract them

not feasible to split offices, were found generally to prefer moving within London from the centre to the suburbs. In fact many corporate head offices in London are found in the suburbs in office centres such as Croydon or slightly further out (e.g. BOC in Surrey and Smith Klein Beecham in Brentford). American companies in particular have not found it necessary to locate their head office within the central area. Thus the America fast food chain McDonalds chose to put their head office in East Finchley near the North Circular Road when they entered the British market.

The size of the office workforce in central London has varied widely with the booms and busts of the economy, but there has been no sign of structural change around an average of approximately one million workers. If decentralisation has occurred at a substantial rate over many decades without having much effect on the total then it follows that new employment must have been created to match those who have deserted the capital. Of the six types of central London office users which we identified above, three have experienced long-term growth. These are business services, media and flagships. Foreign banks, for example, a form of financial flagship, have increased from 114 in 1967 to 470 in 1984, a total higher than in any other city in the world. Of the other categories, government and finance have remained fairly static and corporate headquarters have declined. The overall effect has been that the level of office employment, though static over the long-term has increased in the last seven years of the 1990s since the bottom of the depression in 1992. By the end of the decade employment had risen to the point where the capacity of the tube and the railways to carry the number of commuters has come under strain.

But there is also another force at work, the desire of central London users to enjoy modern property at reasonable cost. Compared with other world cities such as New York and Tokyo, London property appears small and obsolete. The two highest rented areas, the City and Mayfair, are full of listed buildings and subject to height restrictions with the result that the development industry is unable to provide the modern offices that are needed in the areas where there is demand. Linked to this is the fact that businesses, like natural organisms, tend either to grow or die. If they grow in central London their rent costs can rise to a high level forcing them to consider decentralising all or part of their operation. The result is a cyclical process whereby new firms are attracted to London, grow until they reach a size when they need to decentralise and then move out to be replaced in turn by other new firms.

We might call this the seed pod model. Seeds such as peas grow in their pods until the pod breaks and the seed falls out. Similarly some newly arrived firms, for example foreign banks, start by locating in high-cost central locations, expand and then move out. As a new firm they are unknown and seek a location where they will have a high level of visibility and be able to make contacts easily. Being small the high cost of rent per square foot of the space they occupy is less important than the fact that the total is bearable. If they succeed they expand, taking on more staff. They become known and the need for a location offering high visibility becomes less. As they expand the high rent cost begins to hurt. They therefore consider moving out, either to a fringe central location or out of central London altogether. As they decentralise so they are replaced by new small firms and the process is repeated. The form of office property in London in fact works to the advantage of the seed pod model. The prime areas of both the West End and the City are full of small office suites suitable to act as pieds-à-terre for new arrivals. Moorgate in the City is full of the plates of exotic foreign banks, while offices in Regent Street are home to serviced office suites or office hotels, as they are sometimes called. Among the older established foreign banks the decentralisation process has already started. Several American banks have moved to modern premises in Canary Wharf while others have moved out of London altogether, for example American Express to Brighton and Chemical Bank to Cardiff (before its merger).

The seed pod model allows the expanding firm to gain the benefits of both larger modern premises and lower rents for the price of giving up a prime location. A good example is the case of Sotheby's as reported by *The Times* on 12 November 1999:

> Sotheby's, the 225 year old auction house and fine art dealer, is drawing up plans to move out of its historic base in New Bond Street, London. The company, which has occupied its Georgian building since 1917, is understood to want to make a move to a more spacious and modern building, which will fit in with the high-tech image it now seeks to portray. Sotheby's sees much of its future growth coming from sothebys. com, its Internet site. The current accommodation, at 34–35 New Bond Street provides 125,000 ft^2 of space but it is considered to be an 'antiquated rabbit warren' by staff at the building. It is seeking a building that can provide up to 400,000 ft^2 while remaining in the West End.
>
> Estate agents believe that Sotheby's will not be able to find any existing building to match its requirement so it will have to look to one of the new developments on the fringe of the West End, such as the proposed schemes at Paddington . . .

The Sotheby's story summarises well the points made in this chapter. They are, first, not an obvious office user, yet they are seeking conventional office space. Second, they are looking to more than double the space occupied from 125,000 up to 400,000 ft^2, and third, this will involve a move from the fashionable area of Bond Street to the unfashionable precincts of Paddington or some similar area where a large enough new development is possible. Finally, although this is not mentioned in the article, it is likely that the new space will not only be more efficient but cheaper than Bond Street. It is interesting, too, that the growing importance of the Internet to Sotheby's does not appear, as might be naively supposed, to have reduced the firm's appetite for office space.

Buildings wear out, become obsolete and have to be renewed in a process which is as inevitable as anything in nature. The seed pod model shows how this process of renewal is often accompanied by expansion and decentralisation.

The skyscraper puzzle

Offices have generally been discouraged or ignored by planning authorities, the feeling being that work that took place there was somehow parasitic and not real like manufacturing. Offices were permitted, if at all, only in town centres. During the 1980s there was growth of 'hi-tech' developments with a high proportion of offices. These were permitted out-of-town on new parks on the grounds that this would encourage modern forms of industry, though much of the demand in fact came from what would now be called office users who used the non-office parts of the premises for ancillary activities like printing and packaging. This anomaly ended in 1987 when a new Use Classes Order was introduced. This is a government list of activities where change of use can occur within any category without the need for planning permission. A new B1 category of general business use was introduced which included both offices and genuinely hi-tech manufacturing. The effect was to allow offices to move out of town, and as a result many business parks were developed in the South-East, particularly in the high rented area to the west and south of London. Stockley Park near Heathrow is a good example.

The sudden arrival of offices onto land which would previously have been limited to industrial use raised the question of the form they would take, and the obvious place to look was America where office decentralisation had long been a fact of life. Most American cities have a group of skyscrapers at their core surrounded by low rise development. Indeed the newer the American metropolis the starker the contrast between the soaring skyscrapers and the low rise, low density of the other land uses. The office stars of the last third of the century have been cities such as Atlanta, Denver, Houston and Dallas. A traveller approaching these cities over the flat, open countryside sees what might appear as a bar chart of land values with the bars formed by the steel and glass of the city centre skyscrapers.

By British standards, planning controls in these American growth cities is weak and a free market in land more nearly operates. As we showed in chapter 4, high demand produces high rent and land values which in turn lead to the desire to increase the income generating capacity of the land and hence the building of skyscrapers. Very high skyscrapers in the middle of the prairie would seem to confirm the strength of the demand by clustering office users which was discussed above. The puzzle lies in the extremity of the contrast between the skyscrapers and everything else, and the resilience of the decentralised offices. These new cities are built to low density to accommodate widespread car ownership. As a result, virtually every human activity tends to prefer low rise buildings surrounded by ample car parking. Shopping, motels, factories, warehouses, hi-tech, leisure activities, hospitals and colleges, all are decentralised in low density developments and low rise buildings. In fact a majority of office users choose to locate this way too.

The 1980s and 1990s have seen the majority of new development switch from the central business district (CBD) to the suburbs, and rents between the two converge. The American example, then, seemed to point to a low density solution for all land uses except a minority of office users. Except for a handful of the largest and most European of American cities such as Boston and San Francisco, American CBDs have been deserted by all uses except offices and those services which complement them such as hotels, conference facilities and speciality shopping. The lesson for Britain, therefore, would seem to point in the same direction. The problems Britain has, however, are land values over twice the American level and a limited number of sites suitable for campus office development. We need to return to America for another look.

In several American cities low rise, low density suburban offices have been redeveloped as offices with a height of 20–5 floors, often served by multi-storey parking. Then in the 1980s advantage began to be taken of the high accessibility of the suburban junctions of radial freeways and M25-type beltways. These junctions naturally became attractive for regional shopping centres and land values rose rapidly. As a result in the more affluent suburbs they became in effect new town centres. They were, however, unlike any town centre ever seen before. Their form was loose knit and they stretch for several miles along the corridor quite unlike the British suburban centre or the traditional American down-town. Examples are Tysons Corner in Washington, DC, Post Oak in Houston, Costa Mesa in Orange County near Los Angeles and the north-western corridor in Atlanta (see Garreau for a study of this phenomenon). These locations have become attractive for offices. Interestingly, the presence of shopping and other facilities for office workers are valued as well as access to the beltway, the airport and nearby high income housing. Public transport is minimal and parking provision allows for everyone to drive to work. Parking standards are four generously laid out spaces per thousand square feet of office space. Decked parking is common and, unlike Britain, is considered an asset, offering as it does protection from the sun and the snow.

The north-western corner of Atlanta provides a good example of this phenomenon. Atlanta has seen suburban growth as rapid as any American city over recent years, with office stock increasing 40 per cent between 1983 and 1987. The north-western corridor is centred on the junction of Perimeter Highway (the beltway) and the radial Freeway 75. In 1987 it contained 11.6 million ft^2 (1.1 million m^2) of offices in 67 developments compared with down-town's 11.1 million ft^2 (1 million m^2). There are eight office blocks of a height of 16–25 floors, all built since 1981 and all grouped around the intersection and Cumberland Mall, the nearby regional shopping centre. The changing height of the developments since 1981 is shown in Table 9.2. The tenants of these new high offices in the suburbs are mainly corporate headquarters and regional offices of big companies. They moved there because compared with down-town there would be easier access and

Table 9.2 The number of new office developments by
        height in Atlanta, north-west corridor

Number of floors	Pre 1981	%	1981–87	%
1 and 2	16	61	8	16
3 to 9	10	38	27	55
10+	0	0	14	28
Total	26		49	

Source: Landauer, *Estates Gazette*, 1988

parking, less danger of crime and a shorter commute for senior staff. A block of 20 floors forms a prominent landmark which may stand out more than 40 floors in the city centre. It also offers easier access to the airport than down-town. Finally, the presence of good shopping facilities is considered to be a major attraction and widely used in promotion. The result is rising land values and the formation of a new form of town centre designed for the car-using age.

The skyscraper puzzle is in two parts. The first is the acute contrast between skyscrapers in the CBD and the low rise development elsewhere. The second part suggests that the desire to gain the benefits of clustering continues after decentralisation and leads to the creation of a new form of centre where rising land values once again cause development densities to increase.

In Britain it is not possible to point to an exactly similar process happening, partly because there is a strong antipathy to high rise development. Out-of-town office rents have risen in the more favoured parts of the South-East such as the Thames Valley to a level where they are higher than the majority of suburban centres in greater London. The old pattern of London forming a rental mountain standing above the rest of the country has broken down with the emergence of this plateau to the west, roughly bounded by Oxford, Basingstoke, Guildford and Slough. As with shop rents, the gap in office rents between town centres and business parks has largely disappeared. The factors which attract decentralised office tenants are the same as in America, namely a nearby high income residential area, a pleasant environment with good access to facilities such as shopping and good transport links to Heathrow and central London. Of these the most interesting is the desire to locate in high income residential areas. This has led to an imbalance between east and west London with the western half of London and the South-East benefiting in terms of development pressure and land values compared with the east. Like other land uses, offices, once liberated from the old constraints, are free to choose an attractive environment. The trouble is that in doing so they may diminish the very attractions they seek.

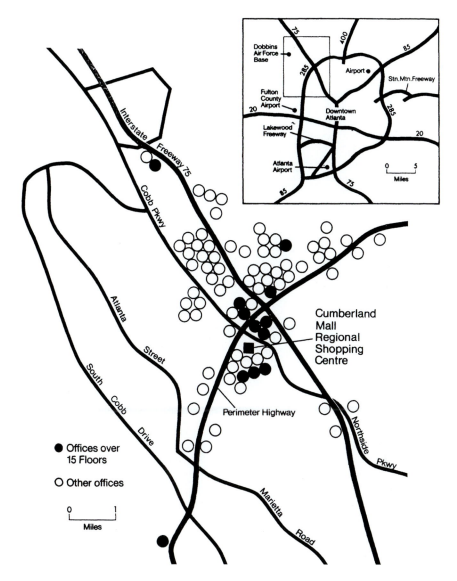

Figure 9.2 North-west Atlanta
Source: Landauer (1988)

Rent contours

In the previous section we looked to America for understanding of non-central office location. Britain, however, is a European country and the situation there is different in one crucial respect. Much of the centres of

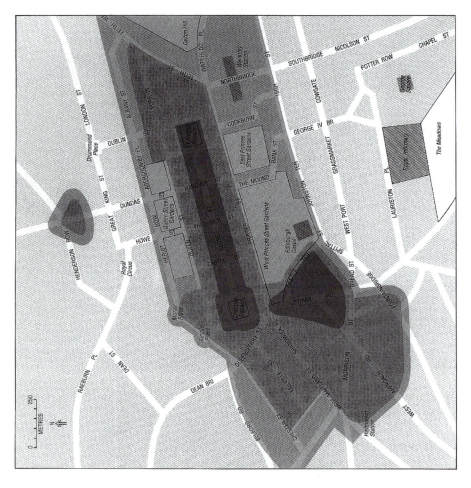

KEY

Rents per sq ft per annum

£
20+
16 - 20
12 - 16

Prime Rent	—
Prime Rent £/sq.ft./p.a.	22
Prime Yield	5.75%
DATAR Score	49

Figure 9.3 Office rent contour map: Edinburgh
Source: Hillier Parker (1990)

European cities are historic and unsuitable for modern offices. Modern offices need to be air-conditioned and to contain large floor plates of 30,000–40,000 ft² (3,000–4,000 m²) and it is difficult to provide either of these requirements in an area of historic buildings. Canary Wharf is remarkable not so much for its height as for its width. Although it is only marginally higher than the NatWest Tower, it is several times as broad. Indeed the size of individual floors of the NatWest Tower in the City and Centre Point at the eastern end of Oxford Street are considered major disadvantages.

The problem the European city has in accommodating modern offices is that the area of high values, which as we saw in chapter 4 is the area that naturally attracts new development, is also the historic heart where such accommodation is difficult. It is difficult because of height restriction and conservation arguments, but also because historic street layouts and land ownership makes the assembly of a large site nearly impossible. Thus while modern skyscrapers can be found not just in America but throughout the world, they are lacking in Europe. The skylines of cities stretching from Vancouver to Sydney and from São Paulo to Kuala Lumpur, Singapore and Hong Kong resemble American cities, whereas in Europe only London and Frankfurt have groups of high rise offices within their traditional centres, and they are modest by world standards.

European cities have reacted to the dilemma of areas of high value being unable to accommodate modern offices in a number of ways and this has affected the contours of rental values. An examination of these contours suggests that three different patterns have emerged. First comes a group of cities where the city centre corresponds fairly closely to the area of highest value. Many of these cities have populations of under a million, such as Edinburgh and Zurich, and have been able to accommodate a significant office function within their historic core. Most of the provincial French and British cities are found in this group, as are Turin and Stuttgart. In Edinburgh it has proved possible to expand the office stock in the sensitive city centre at a steady rate and combine this with larger offices in the suburbs. Some Scottish financial institutions, for example, have both a head office in the New Town and a large back office in the suburbs (Figure 9.3). Three of the largest cities also fall into this group. In London and Paris, modern development is taking place away from the area of highest value, in Docklands and La Defence, and the same is true of Milan, where the contrast between high values in the city centre and dispersed modern development in low value peripheral areas is as marked as anywhere in Europe.

The second group are cities where the highest values include not only the traditional centre but also extend to include a neighbouring high income residential area where the newer development is taking place. This has occurred in Frankfurt where its new skyscraper, the Messeturm lies in this north-western extension while the twin towers of Deutsche Bank also lie in

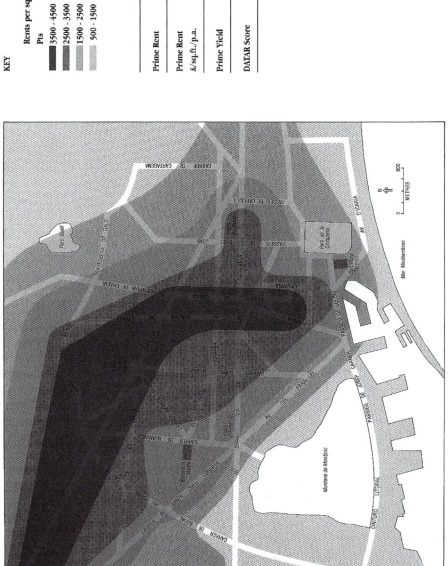

Figure 9.4 Office rent contour map: Barcelona
Source: Hillier Parker (1990)

that direction. Another example is Barcelona where modern development is ballooning out along the Avenida Diagonal and the area of high values is retreating northwards from the waterfront (Figure 9.4). Other examples are Lisbon and Madrid, and Hamburg and Gothenburg in the north of Europe. The advantage of this second group is that the unity of the central area is maintained, but the historic heart is left in peace. This extension is similar to that experienced by many cities in the nineteenth century with the arrival of railways and department stores. Thus London and Paris both extended westward at that time to incorporate high class residential areas (Mayfair and Eighth Arrondissement).

Both cities have found it hard to repeat that experience with modern offices. Hyde Park acts as a barrier to the westward extension of central London as does the bend in the Seine and the Bois de Boulogne in Paris. As a result both London and Paris have had to choose districts for modern offices which are far from both the area of highest rents and the most natural adjacent high class suburb. La Defence (in an old military complex) and Canary Wharf in Docklands are both far away from natural office locations and were selected simply because land was available there.

The third group of cities are those where high values have followed new development out of the traditional centre. The best example of this is Amsterdam where the whole historic city centre lies outside the area of highest value and a new centre has formed well to the south, stretching down towards Schipol Airport (Figure 9.5). In several other cities such as Seville, Nice and Munich new offices have been developed between the city and the airport and it seems likely that before long values there could match or surpass the city centre. Airports are major pulls of attraction for offices, playing the role that railway termini did in the previous century.

According to theory, development should mainly occur in areas of high value, but the evidence suggests that this is often not the case in much of Europe. Decentralisation of offices is occurring quite rapidly throughout the continent, but the movement of value has lagged behind. This could lead to a split between centralised users staying in the city centre and others decentralising to the suburbs, as already described for central London. If this happens the very lack of office space in the centre could ensure that demand remains unsatisfied and rents stay high. In other areas such as Canary Wharf where supply more nearly matches demand, this balance could itself keep rents low.

It is interesting in this context that Mayfair and St James's has now over-taken the City as the area of highest rents in Europe. This area consists mainly of property which was built as aristocratic town houses before conversion to office use. It is unsuitable for firms like Sotheby's or a financial or business service firm seeking 100,000 ft^2 (10,000 m^2) of modern offices, but it can satisfy some office users. Corporate headquarters and flagships may for example find a converted town house ideal, and these,

KEY

Rents per sq m per annum

Dfl

- 325+
- 275 - 325
- 225 - 275
- 175 - 225

Prime Rent	350
Prime Rent	
£/sq.ft./p.a.	10
Prime Yield	6.00%
DATAR Score	63

Figure 9.5 Office rent contour map: Amsterdam
Source: Hillier Parker (1990)

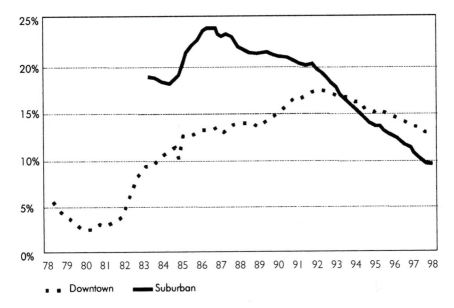

Figure 9.6 American office vacancy rates: downtown and suburban 1978–1998
Source: CB Commercial RCG

together with smaller financial and business service firms are in fact the main occupiers of this privileged area.

It is still uncertain whether top rents and modern development can remain divorced between different locations. In the long term it seems likely that values will follow modern developments out of the city centre. If this happens it will ease development pressure on historic areas leaving them for tourism and residential use, but it could also lead to problems of long-term financial viability. In this respect decentralisation of offices presents the same dilemma as it does for shops.

10 Industrial location

Manufacturing location

It was argued in the last chapter that the majority of work took place in offices, with manufacturing taking a minor share of the economy. One might expect therefore that the section of the book dealing with it to be shorter than that covering offices, which indeed it is. There are, however, a few interesting points worth making. Models of industrial location derive from Weber at the beginning of the century and are based on the idea of cost minimisation. The preferred location was found to be that which minimised the cost of supplying the finished product. This involved taking into account the assembly of raw materials, transport, energy and bringing the finished goods to market. Land and labour were also important. All these costs were combined and the location chosen where they were minimised.

Since the early days of the twentieth century, the movement from heavy to light industry has reached the point where this model is of little relevance. Improvement in transport and the globalisation of the markets for raw materials means that these costs vary little between different locations in Britain. When the northern parts of the UK compete for an incoming factory, say a Japanese car manufacturer, they do so primarily on the existence of suitable large sites, the availability of cheap, good quality labour, and good access to ports. Provided these requirements are met, the difference in costs between say Sunderland (Nissan) and Derby (Toyota) appears to be of insufficient importance to affect the location decision. This is particularly the case when Britain competes with other countries by offering grants and other subsidies.

Globalisation has meant that British labour has had to compete with third world countries where labour is a fraction of the cost. This has led Marks & Spencer, for example, to reduce the proportion of the clothing it buys from the British textile industry from over a half to barely a third. Instead it finds it cheaper to service the British market with clothes made in Asia. Despite the fact that fashion demands relatively fast delivery, it still pays Marks & Spencer to access its clothing from the other side of the world. Other costs are clearly trivial in comparison. With the decline of heavy

Table 10.1 Industrial rental values, 1965–1999

	South-East[1]	North-West
Index (1977 = 100)		
1965 May	31	30
1981 May	182	164
1991 May	480	375
1999 August	445	342
Increase %		
1965–81	487	447
1981–91	164	129
1991–99	−7.3	−8.8

Source: Hillier Parker Rent Index

Note
1 excluding Greater London

industry and the coming of globalisation, manufacturing has declined in the northern half of the country and the population has declined too in favour of the thriving south. In response the government has encouraged new factories to locate in the north, thus accentuating the division between a manufacturing-dominated north and a service-dominated south. The result has been a widening of the gap in prosperity between the two halves of the country.

Hillier Parker publishes an index of rental values dating back to 1965 and this shows the growing gap in rental levels of prime industrial property between north and south (see Table 10.1). Interestingly, the figures show no clear pattern for shops and office rents. There appears to be a growing polarisation between London and its supporting region (the area roughly within a hundred miles of the metropolis) and the rest of the country. With increasing competition from cheap foreign labour hitting manufacturing this gap looks set to widen.

This phenomenon is not limited to Britain, but is found elsewhere. In America there is a contrast between a declining rustbelt in the north and a growing sunbelt in the south, and the same north–south divide can be found in France. In both countries the declining north was nearer to the major centres of population than the burgeoning south. The south seems to have been selected because of a better climate and environment. Once most buildings have air conditioning, a hot climate ceases to be unpleasant in the heat of summer and eliminates the discomfort of winter. The rustbelt–sunbelt split was strongest in the 1980s. The 1990s have seen signs of recovery in the north.

The three factors which are associated with growth industries are attractive environment and climate, a large pool of specialist labour and (least important) proximity to major markets. It should be pointed out here that modern manufacturing involves less metal-bashing than before in the mix of services provided. The ratio of hardware to software has steadily fallen. IBM, for example, makes computers and is therefore a manufacturer, but out of some 20 properties, only two are actually engaged in producing hardware. The principal factory is in Greenock on the Clyde estuary, a location far from the south of England. The remaining plants offer customer servicing including customer training and are distributed to match the location of the customers themselves. Studies of branch closure following merger or major reorganisation show that, other things being equal, it is the more remote plants which are the most vulnerable. Apart from this, however, a location such as IBM at Greenock functions adequately.

The pool of labour is important in areas of new technology such as Silicon Valley near San Francisco, where a concentration of like-minded people in the same physical space act like yeast in causing the productivity of the whole to grow. In Britain a similar effect is found in the motor racing industry. Concentrated in an arc stretching from Surrey to Northants, Britain contains the greatest concentration of specialist motor racing firms in the world. The fact that these firms are within commuting distance of each other is clearly of benefit to employers and employees alike. It is worth noting that neither IBM nor the motor racing industry comprise manufacturing in the generally understood sense of the word. Mass production is missing as are economies of scale. The proportion of the value of the finished product taken by the actual hardware itself is remarkably small, and the ratio of brawn to brain is tiny. What this means is that the division between manufacturing and services blurs to the point that it becomes meaningless. The distinction between blue and while collar workers is disappearing and the type of manufacturing as described in the text books is withering away.

Warehouse location

The extraordinary revolution which has occurred in distribution was described in chapter 3. There we saw how the pallet, the steel container, the fork lift truck and the computer combined to shorten the supply chain and caused huge areas of industrial land to be released for other uses. In this section we look at what this has meant in terms of warehouse location. As with so many recent changes, the trend has been towards fewer but bigger warehouses. This is a trend which is usually accompanied by overall expansion; with warehouses, however, the evidence suggests that the total floorspace is falling. Although consumption generally is rising steadily, this has occurred at a rate slower than the increase in productivity.

The distribution revolution has led to the arrival of the high bay warehouse in the last 15 years of the century (Table 10.2). High bay warehouses are defined as being of a floor area of at least 100,000 ft^2 (9,300 m^2), with the building occupying no more than 45 per cent of the site (called the site cover). What distinguishes them from other buildings, and particularly from

Table 10.2 Warehouse floorspace in England:
estimates for 1994

	Million ft²
High Bay	18
Supermarkets*	15
Logistic contractors*	50
Modern warehouses	100
All warehouses	1500

Source: Valuation Office: Commercial and Industrial
Floorspace Statistics England 1994, HMSO and
author's own estimates.

Note
*including part which is high bay.

other warehouses is their height, hence the name 'high bay'. To qualify they
need an eaves height of at least 30 ft (10 m), though many are 40–50 ft or
higher. The eaves height is the distance from floor to the lowest part of the
ceiling, usually at the wall. The interior height of these new warehouses
often means that a three-storey office block could fit inside. It gives them an
awe-inspiring cathedral-like scale. The height enables them to contain many
layers of racking, sometimes up to seven or eight layers which effectively
more than doubles the former capacity. Man riser fork lift trucks can store
and access pallets, one on top of the other to the limit of the racking. Of
course to do this it is necessary for the floors to be able to withstand great
weight and be of billiard table flatness. For the food and drink industry these
warehouses are air conditioned, being kept at the necessary temperature,
either frozen, chilled or ambient (normal). The result is that although
basically a large shed, the high bay warehouse is in fact a sophisticated piece
of construction.

In chapter 3 we saw how half Britain's food was handled by less than a
hundred supermarket warehouses. Such concentration has also occurred in
non-food areas. Whereas food needs to move fast to stay fresh thus causing
the major supermarket chains to operate from more than a dozen
warehouses spread across the country, outside the food and drink industry
it is often possible to service the entire country from a single warehouse.
There is a difference in size with food and drink warehouses averaging
over 300,000 ft² (28,000 m²) while non-food warehouses average only
165,000 ft² (15,000 m²). In locational terms there is also a marked
difference. Food and drink warehouses tend to be distributed in line with
population whereas non-food warehouses concentrate in one particular
area from where they can service the whole country, namely the South
Midlands.

It is noticeable that supermarket warehouses avoid high value areas such
as London. Instead they serve the London and South-East market from

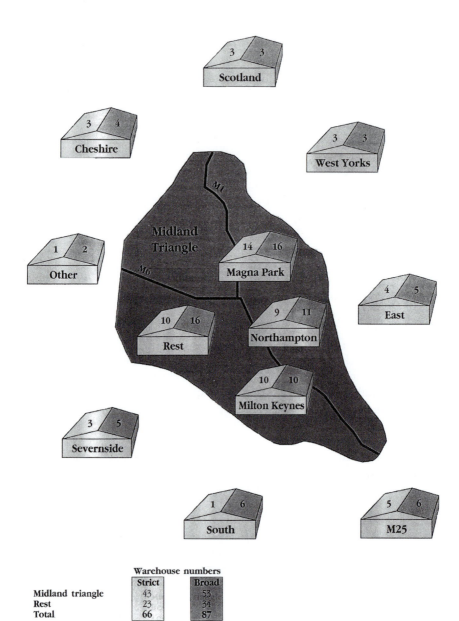

Warehouse numbers		
	Strict	Broad
Midland triangle	43	53
Rest	23	34
Total	66	87

Figure 10.1 High bay warehouse location
Source: CB Hillier Parker (1994)

areas outside the M25 but with good road location Thus J. Sainsbury has warehouses in Hertfordshire at Buntingford and London Colney while Tesco has a giant warehouse to the south of Hatfield. It should be mentioned here that not all the supermarket warehouses are high bay. Some, such as the two serving J. Sainsbury just mentioned, have a traditional low eaves height.

Supermarkets and some other large warehouses tend to cluster in areas of high accessibility and modest rents. Thus there is a concentration to the south east of Glasgow, and at Wakefield and Warrington in the north of England. Increasingly high bay warehouses are found in purpose-built distribution parks with good access to motorway junctions. Wakefield, for example, is near the junction of the M1 and the M62, with the A1(M) not far away. While Warrington, being situated between Liverpool and Manchester, is served by the M6, the M62 and the M56. There are similar concentrations near the Severn Bridge to serve South Wales and the South-West and smaller groups near Huntingdon (A1 and A14) and the eastern half of the M25 in Kent, Essex and Herts.

These concentrations, however, are all dwarfed in scale when compared with the South Midlands area around the junction of the M1 with the M6/A14. Within half an hour's lorry drive of this junction lies what the property industry refers to as the Midland Triangle, an area lying between Leicester, Birmingham and Milton Keynes. Around 40 per cent of the 160 or so high bay warehouses in the country lie in this area, but it contains 52 per cent of all non supermarket high bay warehouses. The Midland Triangle includes the Brackmills Estate in Northampton, the Tongwell and Kingston Estates in Milton Keynes, together with Mid Point in Minworth and Swift Point at Rugby. The giant and pioneer of distribution parks, however, is Magna Park near Lutterworth on the A6. Built on an old airfield, Magna Park contained some dozen high bay warehouses by the mid-1990s.

Magna Park and the other distribution parks in the Midland Triangle serve manufacturers in particular. Apart from the food–non food division, there is also a split between the warehouses of retailers, those of professional logistics contractors, and manufacturers. Manufacturers were the last of these to start using high bay warehouses and they have found that it is possible to serve the entire country from a single high bay warehouse in the Midland Triangle. Indeed Milton Keynes, Northampton and Magna Park could compete for the title of Import City. The list of their warehouse occupiers contains most of the leading foreign manufacturers of consumer goods, with Japanese companies to the fore.

The Toyota story

Perhaps the best way to illustrate the operation of a national distribution service from a single Midland Triangle warehouse is by means of a story. The Japanese car manufacturer, Toyota, operates through 270 dealers in Britain.

These dealers need 100,000 parts of all the various Toyota models, ranging from door panels down to the smallest nut. Every time a new model is produced it involves an extra 5,000 lines. Toyota's customers expect to have these parts available at short notice at each of the 270 dealers. This is undoubtedly a demanding logistical and distribution task, yet Toyota manages to achieve this from a single warehouse. One warehouse of 195,000 ft^2 (19,000 m^2) in Magna Park is sufficient to service the entire country. The warehouse is manned by two shifts of 24 people each.

To illustrate the efficiency of their distribution system, Toyota tell the story of a customer who accidentally locked his car keys inside his own car at a motorway service station. He telephoned his dealer who telephoned Magna Park who in turn were able to despatch a motor cycle courier with the replacement keys so that the entire operation took barely two hours.

There is a final twist to the Toyota distribution story which further illustrates the nature of modern distribution. Toyota operate a manufacturing plant near Derby, not 50 miles from Magna Park. No parts, however, are shipped direct from Derby to Magna Park as one might expect. Rather parts are sent from Derby to a giant European distribution centre near Ghent in Belgium. Lorries then deliver a selective range of parts to Magna Park so that inventory is maintained at the desired level. The Toyota warehouse in Magna Park receives parts direct from component manufacturers, but the more bulky and slower moving lines which are actually produced in its factories are delivered along with other requirements direct from Ghent. By operating in this way it is easier to control the stock at Magna Park at the precise level needed. Delivery operates on a 'just in time' basis so that no more stock is carried in the warehouse than is necessary to service customers' needs.

The Toyota story illustrates the characteristics of modern distribution. The supply line is shortened so that buffer storage is greatly reduced. The steps of the distribution chain are reduced and simplified, and outsiders are excluded. In the motor industry many of the 100,000 parts are made by independent component manufacturers and it would be tempting but less efficient to require them to deliver direct to the 270 dealers from their own warehouses. Instead because the whole country is within a day's drive of Magna Park, it is possible to control the entire operation from a single automated warehouse. For the customer to receive his keys within two hours it is necessary to push the responsibility back onto the component manufacture and this effects the manufacturer's own location decision. For manufacturers to deliver 'just in time' it is easiest if they are near to their customers. Road congestion is now so bad that to ensure punctual, reliable delivery, lorries often travel very early in the morning or at night and park near their destination. It is one of the many pressures pointing towards a 24-hour business day, but it ensures the flexibility necessary for the customer-driven distribution system to operate with the improvement in productivity and choice which goes with it.

Science parks and hi-tech

In the 1970s there was considerable interest in the growth of computer based high-technology industry. Interest centred on the idea of technology transfer whereby university-based research, particularly in electronics and biotechnology, could be transferred to industry and turned into economic growth. The model was California, where the electronic engineering department of Stanford University had spawned a whole shoal of new companies nearby. It was at Stanford, south of San Francisco, that Hewlett met Packard and set up in business in what was to become known as Silicon Valley. The success of Silicon Valley was due as much to the attitude and calibre of the teaching staff at Stanford as to the brains and enterprise of their students. Interchange between the university and the surrounding small companies was encouraged to mutual benefit. The graduates in their fledgling businesses benefited from technical advice and possibly even the use of the university's equipment. In return they offered the university an opportunity for the latest ideas to be tested in open market conditions.

The transfer of technology from university laboratory to commercial business was able to occur because of the synergy between like-minded people and it was the presence of substantial numbers of like-minded people which permitted this synergy to happen. There was, in other words, a minimum scale of operation before take-off could occur. Forty years after the beginning of Silicon Valley the area is still at the forefront of computer technology and has been envied and copied throughout the world. It would be no exaggeration to say that the spin off and employment growth which has followed the success of Silicon Valley has made a major contribution to the American economy and kept it at the forefront of developments in computers and information technology.

Many attempts have been made to replicate the phenomenon of Silicon Valley and most have failed. In America itself hi-tech growth has successfully occurred in a handful of locations. The beltway around Boston is one such place with the role of Stanford University taken by MIT and Harvard. The surroundings of Washington DC is another place as is Houston, both stimulated by the defence needs of government. These three areas appear to have nothing in common with Silicon Valley in location except perhaps that they all lie outside but not too far from the three giant American cities of New York, Chicago and Los Angeles.

In Britain a movement to build science parks began in the 1970s to encourage the imitation of the Silicon Valley model. Mostly they consisted of groups of small starter units and were situated on or near university campuses. By the mid-1980s there were 24 such schemes being developed or complete, and of the space available 69 per cent was occupied. In total these parks contained 1.5 million ft^2 (140,000 m^2) of space. The money spent on these science parks was invested for reasons other than the normal money-making objectives of property investment. Much of it was public money and it was intended to encourage technology transfer between academia and the private sector. The result was that the effort was dispersed in small pockets with the outcome that synergy was difficult. In fact it is worth noting that Silicon Valley grew without any particular property assistance of public sector help, as did the three other American hi-tech areas mentioned. The two British science parks which have proved to be the most successful are Cambridge and Reading. The Cambridge science park which is by far the biggest in the country was developed by Trinity College on their own land on the northern edge of the city. Reading, with its concentration on the food industry, is not usually considered a science park at all since there is no building or development of that name. Reading, however, has developed successful relationships with major food and pharmaceutical companies which occupy premises near the Whiteknights campus of the university.

In 1986 John Henneberry completed a survey of science parks and compared them with 45 hi-tech estates which had been built to normal property development criteria and also took a control group of normal industrial estates. The science parks can claim to be a success in that 68 per cent of their tenants are in hi-tech industries against 44 per cent for the hi-tech schemes and 24 per cent of the control groups. For both science parks and hi-tech estates between 50 per cent and 60 per cent of the space was in office use, a point referred to in the previous chapter, compared with only 17 per cent for the conventional estates (see Henneberry 1987).

The distinction between science parks and hi-tech estates is not a rigid one. Science parks tend to result from public sector initiatives and to be concerned with technology transfer. They are usually situated on university land and consist of small incubator units. Hi-tech estates are normal commercial developments aimed at occupiers in hi-tech industry. They are larger than science parks with a larger office content than conventional industrial estates.

What is perhaps most important from the locational point of view is the difference in the way the different types of development are distributed across the country. As we saw for offices, there is a distinct area of high rents lying to the west of London. This is also the main concentration of hi-tech development. The one exception to this is Cambridge, where the pre-eminence of the university in science has led to something resembling the Silicon Valley phenomenon. The regional spread of the three types of development is shown in Table 10.3. This shows that science parks are distributed disproportionately away from the South-East, while the hi-tech estates are concentrated heavily in the Thames Valley, defined as the western half of the South-East. The science parks, built with a missionary objective, have ended up being distributed in exactly the opposite way to market forces. If Britain is to have hi-tech areas offering the synergy found in America, it seems clear that they are to be found in Cambridge and the Thames Valley. So far science park development has not assisted in this, although it has helped in other ways.

Table 10.3 The spatial distribution of science parks and hi-tech estates

	Science parks %	Hi-tech estates %	Conventional estates %
Thames Valley[1]	17	53	36
Rest	83	47	64
	100	100	100
South[2]	33	67	57
North[3]	67	33	43
	100	100	100

Source: CB Hillier Parker and J. Henneberry

Notes
1 London and the South East west of a line running south down the A1 and continuing to Brighton.
2 South East, South West and East Anglia
3 Rest of UK

Rent contours

Industrial rents are less sensitive to changes in location than rents for offices or shops. Nevertheless there is still considerable variation, with rents for a standard good quality industrial shed in the most expensive areas standing at three times the lowest level. As with offices the overall pattern is for rents to fall with distance from London. Unlike offices, however, the area of highest rents is not central London itself. Indeed standard industrial sheds are lacking there. The area of highest rents, in fact, does not surround central London, as might be expected, servicing the needs of wealthy London occupiers for printing and other services. Instead it lies to the west stretching from the edge of central London, along the M4 to beyond Heathrow. The centre of London has been moving westwards for hundreds of years and throughout the twentieth century modern industry has moved from Hammersmith to Brentford and then on to the Bath Road and so out towards Heathrow and Slough. As the high rent area is now centred on Heathrow it would be tempting to conclude that the presence of the airport was the main causal factor, but there are reasons to doubt this.

True, there is also a high rent tongue stretching from London to Gatwick, but it is not possible to spot a similar pattern at Stansted or Luton, or indeed at Manchester and Birmingham airports. Airports are certainly desirable places to be near, with their demand for in-flight meals and other services and the opportunity for air freight. On the other hand the development of modern industry in west London pre-dates Heathrow, so it is probably best to remain sceptical until more evidence emerges.

Figure 10.2 shows a cross section of rents running from Carlisle to Dover by way of London and Birmingham. This line misses out the high rent area

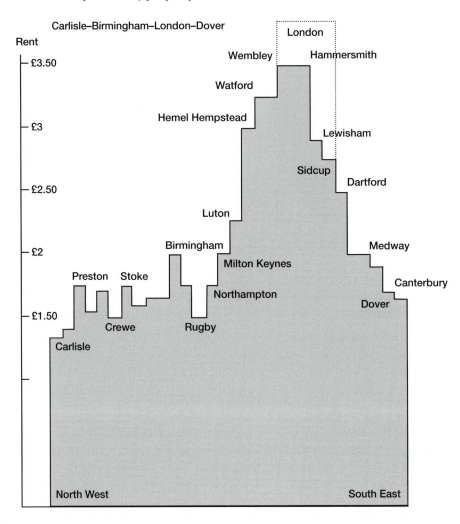

Figure 10.2 Industrial rent contours, 1980
Source: CB Hillier Parker

to the west which is shown in Figure 10.3. The national cross section reveals a number of interesting points. Particularly marked is the low figure for Dover. With the growing importance of Europe as a trading partner and the increase in roll-on roll-off lorry freight through Dover and the Channel Tunnel, one might expect high industrial rents in Dover and the rest of Kent. Instead they are remarkably low. Ports in general have low rents. Thus rents in Liverpool are below Warrington, Hull below York and Felixstowe below Colchester. Rents are determined by the balance of supply and demand, and it could be that the contraction of demand following the distribution

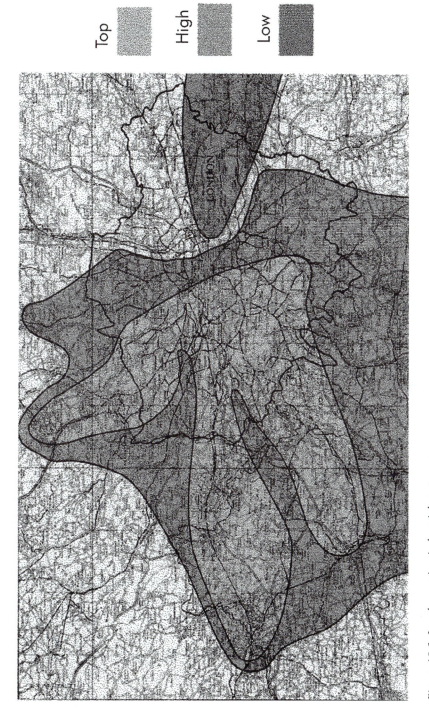

Top

High

Low

Figure 10.3 London region industrial rent contours
Source: CB Hillier Parker

revolution described above is the cause. Certainly the importance of ports as points of trans-shipment is not reflected in the level of rents. The evidence of ports also throws doubt on the importance of airports in determining rents. Airports after all are a modern form of port.

As with ports, new towns might be expected to be associated with high rents, but the evidence does not support this. Although they have specially designed modern industrial areas, new towns in general experience rental levels which are similar to those of neighbouring towns. Undoubtedly new towns have absorbed a substantial part of the demand by new industrial occupiers, but it seems that high supply has kept pace with demand to prevent rents rising.

Another factor which might confidently be predicted to be associated with high rents is proximity to a motorway. Advertisements for industrial property in the property press almost always refer to the distance to the nearest motorway as a plus point, and developers certainly seek sites with as good motorway access as possible. The evidence from the rent contour map, however, is inconclusive. There are certainly tongues of high rents stretching down the M3, the M40 and particularly along the M4, but it is hard to see a similar pattern elsewhere. There is no equivalent tongue, for example, along the M1. It is also worth noting that cities which are some distance from motorways, for example Norwich, Bournemouth, Plymouth and Aberdeen, all have relatively high levels of industrial rents. There is no doubt that motorway access is important in industrial location, but as with new towns, it is likely that supply has balanced demand to prevent rents rising particularly high.

One interesting feature which clearly emerges from the industrial rent contour map is the division between east and west. Reference has been made to the low rents in Kent, and the same applies to Essex and East Anglia. Further north the West Midlands has higher rents than the East Midlands, while North-West rents exceed those in Yorkshire. The most likely explanation for this is that high rents seem to be associated with overall national accessibility. Rents are highest to the west of London forming the central part of Southern England, and high rents generally run along the London–Birmingham–Manchester axis which is the industrial backbone of the country. Generally, rents tend to fall towards the periphery where accessibility is lower. As we saw with high bay warehouses, preferred locations are those that offer access, not so much to a neighbouring conurbation as to a wider national or regional market.

Motorways allow modern lorries to range over most of the country in a single day, even though congestion often means that the journey is undertaken at night. This creates a demand in areas which offer this accessibility which can be strong enough to compete with other land uses, even in the Thames Valley.

11 People and houses

Demographics

The study of where and how people want to live is of fundamental importance in understanding location. The total UK population is growing, but of equal effect is the changing age profile and the move towards smaller households. This in turn leads to different demands on where people want to live, and, with growing car ownership and affluence, greater power to back those demands. In common with most Western countries the population is steadily ageing and the proportion of those beyond working age increasing. This general trend is being exaggerated by the distortion to the birth rate caused by the Second World War. The bulge in births which occurred in the late 1940s reaches retirement age around 2010 and with it will come pressure on traditional retirement areas such as the coast and the South-West peninsula.

The ageing process supports the other fundamental change which is the fall in household size. The average size of household is, on government projections (1992-based), set to fall from 2.67 in 1981 to 2.17 in 2016. This may not sound much but it implies a 23 per cent increase in the housing stock. The implications of this are discussed in the next section of this chapter, but it is not difficult to see the profound effect it is likely to have. Household size is falling for a number of reasons. The birth rate has been falling so that the number of children per household has dropped. The ageing population means that the proportion of households consisting of couples whose children have left home, often called empty nesters, is rising. Finally rising divorce is a further major cause. Unlike the other two factors, however, divorce will probably have only a short-term effect. Around a third of marriages end in separation leading to a demand for separate accommodation, but it is the increase in the growth of divorce which has contributed to the fall in average household size rather than the level itself. It seems unlikely that the rate of marriage breakdown can rise much further, and indeed this is the American experience, so the rate of divorce should stabilise. If this happens it will cease to contribute to the fall in household size.

The changes outlined about suggest that the population is becoming ever more footloose. Increasing numbers of young adults attend universities away from their home town and hence are more likely to move away from their home district on graduation. Similarly the growing number of empty-nesters, often quite affluent, have the freedom and money to move on retirement, particularly, as is increasingly the case, if retirement comes early. These two major factors, plus the generally quickening pace of life, means that people are less tied to any particular locality. Where people choose to move with their greater freedom is a matter of some concern. In the 1960s and 1970s the pattern was quite clear – there was a general move from urban core to urban periphery. In the 1980s the trend continued but was less marked. The population of Greater London, for example, began to rise after a long period of decline although the other conurbations continued to decline at a slower rate. In the 1990s the slow-down noticeable in the 1980s appeared to be accentuated and patterns of movement seemed to be becoming more complex than before.

In America it was noticeable that population growth was occurring in some states, such as Vermont, which lay well beyond commuting range of major cities. Migration in the West began to move away from the magnet of California to undeveloped rural states such as Nevada and New Mexico. People began to talk about counter-urbanisation, a radical shift from urban to rural lifestyles. Whereas before, the movement had been from the city to the surrounding countryside which was within commuting distance, now the move was from urban to distant rural regions well beyond commuting range, and the trend first spotted in America became noticeable in Britain.

It became clear that a major change was taking place. Ever since the Industrial Revolution population had moved from the countryside to the towns. During the nineteenth century this meant a huge shift in the centre of gravity from south to north. Until the second half of the eighteenth century the south with richer agricultural land had been more densely populated than the north. The Industrial Revolution changed that with rapid growth in the towns on the coalfields which were mostly in the northern half of the country. By the early years of the twentieth century this move had run its course. The rural-to-urban trend continued but the destination changed to southern towns such as Slough and Luton and to seaside towns such as Southend. As the century progressed the population centre of gravity swung back to its traditional position in favour of the South (see Table 11.1).

On top of this pattern a second trend was superimposed. This was a move from city to suburbs, from urban core to its surrounding ring. As the table shows, rings have been growing faster than cores at least since 1951, and the trend has continued to 1991. It reached its peak in the 1960s when the cores of large towns grew by only 0.7 per cent while the surrounding commuter ring achieved a massive growth of 17.8 per cent. Since then the rate of growth of the rings has fallen to 9.7 per cent in the 1970s and 5.9 per cent

Table 11.1 Population change, 1951–1991

	1951–61 %	1961–71 %	1971–81 %	1981–91 %
Great Britain	5.0	5.3	0.6	2.5
London	0.4	−4.9	−8.6	1.1
Conurbations	3.4	−3.1	−8.2	−5.1
Rural	−0.6	5.4	8.9	7.8
North	3.3	3.5	−0.4	0.1
South	7.2	7.6	1.9	5.4
Urban cores	4.0	0.7	−4.3	−0.1
Urban rings	10.5	17.8	9.7	5.9
Outer areas	1.7	11.3	10.0	8.9
Rural areas	−0.6	5.4	8.9	7.8

Source: Champion and Dorling (1994)

in the 1980s. The 1970s and 1980s were the decades when growth in the outer and rural areas began to dominate. The outer areas as shown in Figure 11.1 are smaller free-standing towns beyond the influence of the conurbations and cities which are included in the core and ring categories of Table 11.1. As Figure 11.1 shows, outer and rural areas have reflected the greater growth, outstripping both the cores and rings in the 1980s. One can conclude therefore that for the first time in centuries the pattern of net rural to urban migration has come to an end.

Certainly a more detailed examination of rural areas shows that some are still declining in population. Young people continue to leave rural areas for towns with their greater employment opportunities. What has happened is that this trend has been countered by incomers, often affluent people who have caused village house prices to rise, or new houses to be built if permitted.

The incomers can be divided into those who are retired and those still working, by no means a clear distinction. The retired, with their preference for more peaceful areas where they can enjoy the countryside and the sea at lower densities, have been important migrants for decades. What is new is for those working to follow the same paths. The Henley Centre (see Figure 11.2) has reported that when asked what type of environment they would choose to live in, people vote for a rural lifestyle. Today with the benefit of car, modem and satellite dish and with business operating in smaller and more footloose units, they have greater opportunity to realise this desire than at any time before.

When the areas of greatest population growth, both recent and forecast, are mapped they form a ring stretching round London at a radius of around 100 miles. This ring stretches from Dorset and Somerset past Bristol to Northants, Lincolnshire and East Anglia. These counties count as outer

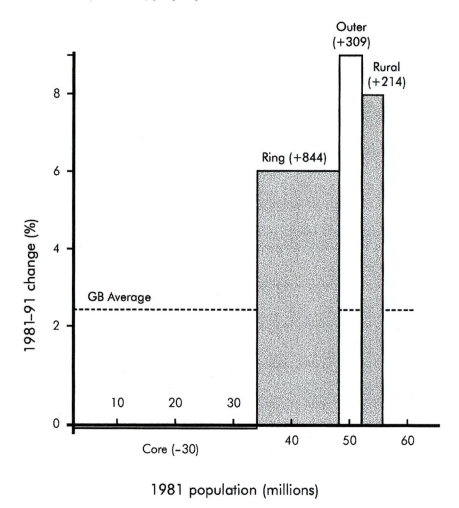

Figure 11.1 Population change by urban core and ring, 1981–1991
Source: Champion and Dorling (1994)

areas or rural in terms of Table 11.1. They top the growth league because
they are at the peak of the ripple of growth which has been spreading out
from London for most of the twentieth century and they also form examples
of the pattern just described. An example of this is West Dorset. This large
rural and coastal district is centred on Dorchester and contains the towns of
Sherborne, Bridport and Lyme Regis. It contains no town of over 25,000
population, and yet during the 1980s it grew by 10 per cent, four times the
national average. It contains no obvious growth industries and is not a
prominent centre for retirement, although some of the population growth

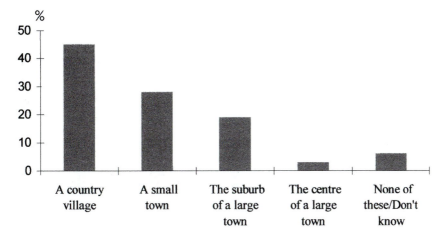

Figure 11.2 Where people choose to live: 'If money, jobs and housing were not a problem which of these would you prefer to live in?'
Source: The Henley Centre

does come from this source. It is reasonably accessible, being just within commuting distance of fast-growing Bournemouth and Poole, and a little over two hours' drive from Heathrow, but it is difficult to point to a cause of its rapid population growth other than its attractive environment. Nevertheless jobs in the area have expanded at a remarkable rate as firms have responded to the allure of coast, Downs and honey-stone villages.

West Dorset is the home of Prince Charles's experimental housing development at Poundbury on the edge of Dorchester. Poundbury is built to a high level of density as an example of how the pressure for new housing might be met. The desire of the footloose to move from the 'madding crowd', to quote Hardy, who lived in Dorchester, is a real sea change in housing pressure and it leads to the issue of housing density which is the subject of the next section.

A question of density

Most British towns follow a similar pattern of density. Surrounding the centre is an area of dense terraced housing. Next comes a belt of semi-detached housing often built in the 1920s and 1930s and beyond that are detached houses at lower density still. This typical structure of rings of declining density fits in with the locational theory which applies to commercial uses and has been touched on a number of times in this book already. Values tend to fall with distance from the centre and as values fall so to does the density of development. As we have seen this generally applies to offices, retail and industrial land uses, although with the coming of widespread car use and hence decentralisation, this pattern was shown to

Table 11.2 Urban and rural living in England, 1991

	Population		Land	
		%	Hectares	%
Urban	42,443,365	90.2	1,086,502	8.3
Rural	4,611,424	9.8	11,956,868	91.7
Total	47,054,789	100	13,043,370	100

Source: OPCS (1997)

be breaking down. It is not surprising therefore to find the same pattern applying to housing, but as with commercial land uses, the old picture is changing.

For commercial land uses it is possible to show that the breakdown is due to natural forces. These forces may contribute in the case of residential density, but the major cause is regulatory action. With widespread political support, the planning system has operated so as to restrict the spread of urban areas into the surrounding countryside. The result has been to restrict the supply of new housing and to increase the price of urban land. The density of buildings in urban areas has increased giving rise to the bizarre sight of high density housing estates being built on the edge of towns so that the density gradient, having fallen steadily with distance from the town centre then rises sharply at the circumference before falling drastically in the surrounding rural areas. The effect is to draw the boundary between urban and rural as tightly as it must have been in the days of medieval walled cities.

There are many different definitions of urban and rural land. The results of that used by the 1991 census of population is shown in Table 11.2. This shows that in England 90.2 per cent of the population is classified as urban but it occupies only 8.3 per cent of the land area. It seems amazing that in one of the most densely populated areas of the developed world such a high proportion of the population should be constrained to live in such a tiny minority of the land, particularly, as we see in Figure 11.2, when the desire for rural living is so clearly expressed. It is a remarkable tribute to the effectiveness of the British planning system. Another illustration of the contrast in density is given in Bibby and Shepherd (1996) who quote Craig's 1987 definition. This shows that urban wards had a density 45 times that of rural wards.

This point was strongly made by Sir Peter Hall in his magisterial thousand page study, *The Containment of Urban England* published in 1973. The constraint he referred to resulted from the 1947 Town and Country Planning Act, which introduced the Green Belt. The Act gave local authorities new power to regulate the use of open land in designated green belts surrounding urban areas. In practice the supply of new housing land failed to keep pace with demand, with the result that house prices rose faster than incomes. In the 1960s the cost of land as a percentage of the housing package rose three-fold. The inevitable result was to increase density and lower the quality of housing the population enjoyed in terms of space. The result has been the development of the high density estates on the outer edges of towns which has become a feature of the English landscape in recent years, and which are sometimes derisively described as consisting of 'rabbit hutches on postage stamps'.

The imbalance between demand and supply which has caused this phenomenon has arisen from the clashing of various trends. Supply has been constrained by the desire to preserve the countryside and protect it from development, but demand has also been increased by the fall in household size described in the previous section. Had it not been for the fall in the size of the average household, the modest rise in total population could have been accommodated without the distortion of the housing market which has occurred.

The fall in household size means that the population of a given urban area with a static number of houses will fall. A town has to grow physically just to stand still in terms of population. This in turn leads to increased pressure for in-filling within the urban envelope. School playing fields are sold for housing, and houses with large gardens are demolished and redeveloped as flats or small detached housing. In terms of the coverage of the urban area by bricks and mortar the density is rising to the point of discomfort, but in terms of people per acre, the density is still falling or barely holding still.

Since 1973 when Sir Peter Hall published his book, the mismatch between supply and demand has got worse. The fall in household size has continued, though as we saw the fall is approaching the bottom. The restraint of allowing housing on 'green' land has also got stronger. In 1988 the government produced a report on the urban land market for the OECD. This stated:

> The rate of urban growth is lower now than at any time since the Second World War. . . . The average annual transfer of farm land to development in England and Wales has fallen dramatically since the 1930s, from around 25,000 hectares per annum to less than 5,000 hectares per annum between 1980 and 1985.
>
> (DoE 1988)

The report continued that in 1986, of the land developed for housing, 46 per cent was 'brown' land, i.e. previously in urban use, and 45 per cent had previously been in agricultural use. Tables 11.3 and 11.4 show various comparisons between the South-East and North-West regions taken from this report. They show in particular how pressure is concentrated in the South-East surrounding London. We will explore this further in the next section.

Table 11.3 Regional land prices

(a) Land prices

	South-East	North-West
1970	21	21
1986	297	131
Increase (times)	14.1	6.3

Note
Index: 1980 = 100

(b) Housing plot prices as % of house prices: weighted average per plot as % of mix-adjusted price of new houses

	South-East	North-West
1970	23	15
1986	38	12

Source: DoE (1988)

Table 11.4 Land prices in 1986 (per hectare)

Regions	Housing	Agriculture	Ratio
South-East	£483,000	£3,600	134:1
North-West	£96,000	£2,890	33:1
Increase (times)	5.0	1.25	

Source: DoE (1988)

Residential location

So far in this book we have seen that values have tended to fall with distance from the town centres. Although the pressures for dispersal and decentralisation brought about by the car have greatly reduced this pattern it still holds good in the majority of cases. We saw how shop rents fall with distance from the prime pitch and how both office and industrial rents decline with distance from London. For commercial and industrial property, therefore, the rental evidence still by and large supports the traditional theory. Commercial land uses usually outbid residential and as a result land values generally follow the same trend. For housing, however, the position is very different.

Whereas for commercial uses decentralisation has tended to even up the difference in value between the centre and periphery, for residential land this pattern does not apply. A number of studies have shown that house prices commonly rise rather than fall with distance from the city centre, for example studies of Leeds and Edinburgh. It is not difficult to explain this. City centres and the surrounding inner city areas tend to be associated in people's minds with factories, congestion, noise, dirt, poverty and crime.

City fringes on the other hand offer more peaceful living and a nearer approximation to living in the country. This as we saw in Figure 11.2, is where a large group of people would choose to live if they had the choice.

In commercial property there is a conflict between the desire to cluster and the desire to disperse. In residential property there is also a conflict between the desire for space and the need for access to jobs and services. The result is that everyone seeking to buy a house or rent a flat has to make a trade-off between space and accessibility. Increasing personal mobility, particularly widespread car ownership, but also the spread of the domestic telephone and now advances in information technology have tipped the balance in favour of the dispersed low density location. Someone living on the outer edge of a city or even in the surrounding countryside suffers very little diminution in accessibility compared with someone living more centrally provided they have ready access to a car and the other services mentioned. Certainly the differential is far less than it used to be. Remoteness has become steadily less of a terror over time.

It may not seem surprising then that the rich should generally prefer to live in these outer areas, and indeed the evidence clearly shows that those well enough off to have the choice tend to prefer low density dispersed living to the high density of the central area. This trend is perhaps not as obvious nor inevitable as it at first appears. There are two points of qualification to bear in mind. First, there is a paradox in the fact that the rich don't live in the areas of the highest land values. In commerce those best able to pay occupy the most expensive locations and one might expect the same in housing. Second, it could be that cultural factors are important here. Not all the rich live at low density, and there are many exceptions to the general rule such as Knightsbridge and Belgravia in Central London. Also the situation is very different in continental Europe.

As a land use residential has to compete with other uses which usually have more financial muscle behind them. The result has been high density residential development on high value land, something which in the Anglo-Saxon mind tends to be thought undesirable, yet high density living need not be unattractive. Central area luxury flats can fetch a very high price as they do in the centre of most of the world's great cities. Where the cultural difference seems to be most marked is in smaller towns and cities. Very few towns in Britain outside London have high-income housing in their central areas. It is possible to name a few exceptions such as Edinburgh and Bath, but even in great cities such as Birmingham and Manchester it is not possible to find the equivalent of Knightsbridge in London. The original high-class inner city neighbourhoods such as Edgbaston in Birmingham and Didsbury in Manchester have changed their original role to commercial and small flats, and most of the original big houses have changed from single family occupation. The pattern is even more marked in America. Apart from New York, possibly Chicago, and a handful of the more European American cities such as San Francisco, Boston and Washington,

Table 11.5 Composition of urban land in
1969 using air photography

Mainly residential	61%
Mainly commercial/industrial	18%
Mainly educational/community	1%
Transport	7%
Urban open space	13%

Source: DoE (1988)

the central areas of American towns are virtually free from high-income housing and indeed many are free of housing of any but the most depressed type.

Britain and America share this Anglo-Saxon attitude in contrast to the Continent and much of the remainder of the world. There appears to be a difference between the type of rich people who prefer to live in the centre and those who choose low density, and this applies also to those of other income groups. People who opt for low density tend to be those who are raising a family while those in the city centre are more likely to be unaccompanied by children and living in small family units, often as a result of divorce. It makes sense that the former group will put a higher value on peace, safety and access to open space while the latter give a greater value to proximity to theatres and restaurants. New York provides a vivid illustration of this division. The American postal area (zip code) with the highest income per head lies in central New York on the Upper East Side. Its 64,624 households contain only 107,000 residents, an extraordinarily low number of less than two people per household.

It is an over simplification to think of cities as forming simple cones of value, certainly in the case of housing (Table 11.5). In many cities as they grew there developed wedge-shaped areas of high quality housing stretching out from the centre. In Bristol an example would be Clifton. It is noticeable that these sectors developed mainly to the south-west of the centre, the direction of the prevailing wind. It has been suggested that during the Industrial Revolution when such areas were formed, it was preferable to be on the windward side of the city to escape the smoke from the factory chimneys.

In London the south-west sector forms the main area of affluent housing stretching from Kensington and Chelsea out through Richmond Park to Esher, Weybridge and beyond to Surrey. There are similar but less marked sectors in the north-west (Pinner and Ruislip), the north-east (Woodford) and the south-east (Bromley). They are separated by what has been called a working-class cross, two lines of factories and poorer quality housing, one running east–west along the line of the Paddington railway and east through the Docklands to Thurrock, and the other running north–south along the line of the A10 and southwards down the line of the A23.

This complicated pattern of alternating sectors may be peculiar to London, but it is possible to see patterns of similar complexity in other great cities such as Paris and New York. It is among the smaller settlements that the basic pattern is clearer to see and it is among such towns that the contrast between the Anglo-Saxon model of Britain and America and the continental European model which applies to much of the rest of the world is most marked. British visitors to France are often astonished to find prosperous and professional families living in central area flats not only in the large cities but in medium-sized towns as well. There are many advantages to the continental model, not least that it leads to lower use of the car to visit services and the place of work. This explains why although car ownership on the Continent is often higher than in Britain, car usage, particularly in urban areas is less. Strange as it may seem, the amount and proportion of trips undertaken on foot are lower in rural areas than in urban, and do in fact increase with density.

The continental model shows that it is quite possible for a significant amount of the high quality housing to take place in areas of high value. All that is required is for people to agree to give up their garden and agree to live in large apartments. With a dearth of new housing land in the country and an increasing number of small households it is possible that Britain will shift more towards the way of the Continent. To help bring this about the slope of land values dropping with distance from the centre needs to become less steep. As we have seen there are signs of this happening and it should continue. It has been argued that land values in the inner cities have stayed artificially high due to market failure. Some derelict industrial sites in such areas are often kept on the books at old values whereas, particularly if they are contaminated, the value in reality may be much less. Some old gas works have even been given negative values to reflect the expenditure needed to bring them back into beneficial use. This plus the lack of new low density housing is pushing this country in the direction of the Continent away from that of America.

While this is happening there appears to be a movement in the opposite direction on the Continent. Decentralisation of population is happening rapidly there following on the British example of some decades earlier. The car is finally having the same corrosive effect on high density living as it has had in Britain and American. It seems likely that in future the Anglo-Saxon and continental models will converge.

Location theory

Evans describes two theories of residential location: filter down and trade-off. Filter down was developed by Burgess in the early 1920s to explain the pattern of residential location in Chicago. Burgess observed that the higher the household income the further out they lived and the newer their housing. Because Chicago was expanding rapidly the wealthy moved out to newer housing further afield, leaving their former homes to 'filter down' to the less well off, while the poorer houses in the centre were demolished to make way for an expanding commercial core.

Trade-off theory replaced trickle-down in the 1960s and 1970s and was developed by Wingo and Alonso and also by Evans himself. In contrast to the supply-side trickle-down, trade-off was demand led. Assuming that land values fall with distance from the centre, the householder 'trades off' minimising travel to work against housing costs. Assuming again that employment is in the city centre, then housing costs fall with distance while travel costs rise.

Evans argues that whether a rich family locates in the centre of a city or on the edge depends on the elasticity of demand for space. If elasticity is high then they will locate on the edge while if it is low they will prefer the more central location. In other words, with a given, albeit high, income, a household badly wanting space will choose the edge while one wanting it less badly will choose the centre, and this conforms with the argument put forward in the previous section.

Trade-off theory provides a reasonably satisfactory model of the broad pattern of house prices, but it operates less well, as Evans admits, for smaller towns or explaining the sectors of alternating values which we examined earlier. In particular the assumption that the person making the trade-off will be someone working in the city centre is increasingly unrealistic. With growing decentralisation of employment the proportion of the population working in the centre is steadily falling and is barely a quarter of the total. The evidence of the journey-to-work tables given in the census show an increasingly complex pattern, not only of movement of jobs from centre to edge but from one town to another. Travel is still an important cost to be taken into account, but with two or more members of a household working in different places, not to mention travel to school, cost minimisation becomes more complex.

Evidence

We end this chapter by looking at three key studies of house prices which throw light on the issues which have been raised. In 1992 the government published research which had been commissioned from Gerald Eve and Cambridge University on the relationship between land supply and house prices using Valuation Office records. The aim was to investigate the extent to which land supply and the operation of the planning system have affected house prices since 1970. The study examined two areas of the South-East, Reigate and Wokingham, and Beverley and Barnsley in Yorkshire. Land supply has been restricted by the planning system, in particular the Green Belt. According to government figures Green Belts now cover 14 per cent of England, more than the total urban area, and have more than doubled in size since 1979. In fact in the South-East, about 40 per cent of the land is subject to some form of protective designation, e.g. Area of Outstanding Natural Beauty, Green Belt or Site of Special Scientific Interest.

The study found that housing land prices 'rose by about twice the rate of house prices between 1968 and 1988, some 350 per cent above inflation'. Constraints on the supply of housing land may have increased house prices in the South East by 35–40 per cent. It also found that the density of detached houses almost doubled in Reigate in the 1960s and there was a further rise of 65 per cent by the late 1980s. Two other findings are perhaps less expected. The first was that the new houses were no more expensive and in fact tended to be cheaper than similar sized second-hand houses. The second related to the substitutability of housing supply. Unlike Yorkshire, there was evidence of similar price movements between locations in the South-East for comparable dwellings. 'Smaller houses in Reigate are close substitutes for similar dwellings in Wokingham.' In other words the report finds that there would need to be a substantial increase in the supply of houses in the South-East for it to have any effect on house prices, and Reigate and Wokingham although lying to the south and west of London, some 40 miles apart, are all part of the same huge housing market. Figure 11.3, taken from the study, shows a development on the edge of Wokingham, near Reading. The plot size is notably smaller than in the older houses which are seen surrounding the development. In particular the gardens are considerably smaller.

The second study deals with the ripple effect of house prices and is by Geoffrey Mean, published in 1997. The idea that the rent and yields of commercial and industrial property ripple out from central London has been touched on before in this book (chapter 4). The evidence suggests that such a trend exists but it is not overwhelming. Mean, on the other hand, finds that 'house prices exhibit a distinct spatial pattern over time, rising first in a cyclical upswing in the South-East and, then, spreading out over the rest of the country'. The post-1990 collapse in house prices was strongly concentrated in the south. As we move northwards, the price fall becomes smaller and in the north and Scotland, no fall occurred at all. There is empirical evidence that the ripple phenomenon has existed at least since 1968.

Mean emphasises the importance of mortgage debt and hence interest rates, followed by earnings and unemployment as factors which are important in influencing the ripple phenomenon. Mortgages tend to be larger in the South-East making the region more vulnerable to interest rate changes. This makes house prices there react sooner and more sharply than elsewhere in the country.

The third and final study is interesting because it covers land prices over both a large area and a long period of time. It presents the results of 749 land sales in the Greater Boston area of America from 1850 to 1989 and covers a 30 mile radius from the city centre. During the near century and a half of the time period covered, the population of Boston grew rapidly and the physical area expanded explosively.

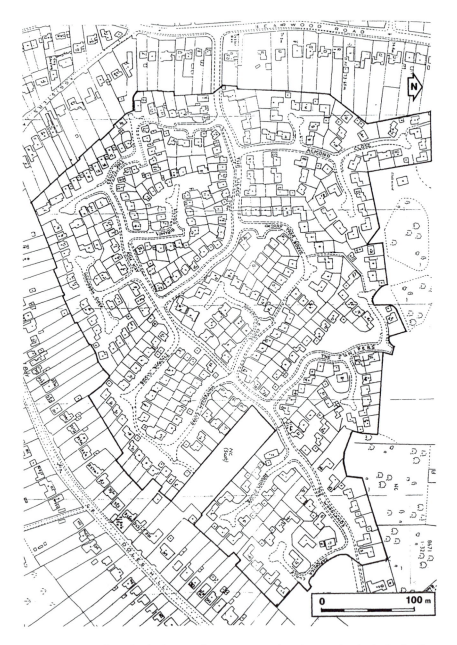

Figure 11.3 Modern high density housing – Wokingham: Elizabeth Park (late 1980s)

Reproduced from the Ordnance Survey map, by permission of Ordnance Survey on behalf of the Controller of Her Majesty's Stationery Office, © Crown Copyright, MC 100033370.

Table 11.6 Land prices in Greater Boston, 1850–1989*

| Year | Miles from Boston | | | | |
	0–5 $	5–10 $	10–20 $	20–30 $	30+
1850	0.71	0.14	0.07	0.04	0.03
1875	3.01	0.99	0.05	0.02	0.01
1900	1.51	1.40	0.07	0.04	0.20
1920	1.73	0.90	0.64	0.30	0.16
1940	3.64	1.24	0.39	0.21	0.44
1960	2.68	1.57	0.66	0.72	0.33
1989	3.36	5.43	0.74	1.63	0.80

Source: Aldridge, Eastman & Waltch *et al.* (1990)

Note
*Pure Land Sales: Median price per square foot in 1988 US dollars

Prices in the inner ring within five miles of the centre grew at a slower rate than any other of the concentric rings within the city region. Land prices there actually fell after 1960 to a level below those of the five to ten mile ring, although they were still more than twice the level of the outer rings. Table 11.6 gives the average price of land through time for five rings of distance from the central area. It shows how the growth of the city reached out to 30 miles by 1920. The prices in the table are very volatile, suggesting that the sample size may be too small to draw detailed conclusions, but the broad pattern appears quite clearly. It is possible to see the reduction in the differential between urban and rural form from about 20 to one in the nineteenth century to nearer five to one by the last quarter of the twentieth century as the city's influence spreads out into its hinterland. It shows also how the rate of increase rises with distance, as might be expected. It should be noted too, that the prices quoted are adjusted for inflation and therefore indicate quite a substantial real increase over a long period.

The three studies quoted here, none of which are widely known, shed light on some of the ideas which have been touched on earlier. All throw a spotlight on land prices which are fundamental in explaining how people choose to live. It is interesting that even in low density, free market America the cost of land has risen ahead of inflation for over a hundred years.

12 Home shopping and the Internet

Introduction

This book has tried to paint a picture of the way location works at the end of the millennium. In doing so it has examined the changes that have occurred over recent decades, such as the increased use of the car, and attempted to place the results into a framework of ideas. It is fitting therefore to finish with a brief chapter on the change which has caused most excitement at the end of the 1990s, namely the coming of the Internet.

The Internet is the culmination of a number of advances in communication which have made the exchange of messages and the sending of data very much cheaper and easier than before. At first sight it would seem that easier communication might have the effect of reducing the need to travel and that people would be able to do more from home or from the office without the nuisance of undertaking a journey in this increasingly congested world. In fact experience suggests that this is unlikely to be the case and previous improvements in communication have coincided with increased rather than reduced travel.

The telephone was invented at the end of the nineteenth century. Together with the telegram, it brought about a revolution in communication which one might argue was at least as revolutionary as the Internet. Until the telephone and telegram, communication had stayed static since before classical times. Messages moved at the speed of a horse. Suddenly with the new invention they were almost instantaneous, and one might naively assume that the ability to hold a conversation over a long distance would reduce the need for face-to-face contact. In fact the invention of the telephone roughly coincided with the invention of the car and later the aeroplane, and travel increased enormously. It is reasonable therefore to be sceptical whether the Internet will reduce the number of journeys. The fact that it is possible to do more from home or from the desk does not mean that people will go out less. Because it is easier to perform routine tasks the result could well be to permit more free time to go out.

Studies of the effect of improvements in communication technology suggest that initially these improvements have a centralising rather than a decentralising effect. This is because the new technology starts in the capital city and the first cable is laid between the two most important city centres. Ultimately the technology becomes widespread and inventions such as the satellite dish and the Internet free people to locate where they want. As we have seen a number of times already in earlier chapters this has the effect of increasing the premium on an attractive environment.

In retailing terms the main effect of the Internet is to divert trade away from existing shops. As we will see later, the effect will vary by trade but forecasts suggest an overall diversion of around six per cent over the next few years with possibly a higher figure in the long term. This is a considerable blow to the town centre but it is unlikely to be disastrous; six per cent is, for example, no more than three or four years natural growth in sales. Also the impact will fall on out-of-town shops as much as those in the town centre. The town centre, or more accurately the traditional retail hierarchy has already lost around a quarter to a third of sales to out-of-town shops. Compared with that the effects of the Internet on the town centre should be modest.

Home shopping

The Internet allows shopping to take place from home. In the excitement which was generated by its arrival, it was easy to forget that home shopping already exists, in fact it has been a significant force in retailing for at least a hundred years. We consider the particular advantages of the Internet in the next section, and these are considerable, but in so far as it allows shopping to take place at home it merely duplicates the several methods of home shopping which have been used for years.

The oldest and biggest type of home shopping is mail order. This involves the shopper buying from a catalogue direct from the supplier either via an agent or through the post. In its modern form it is possible to date the start of mail order back to the days of Sears Roebuck in Chicago at the end of the nineteenth century. There was then in the Mid West a large population of farmers living at low density but with the desire to buy clothing and household goods. Farming was prosperous compared to Europe and operated on a large scale using modern methods and machinery. The result was that there was money earned by the farmers to back their needs for all the new household gadgets which the factories of America were producing. At the same time because of the low population density it was difficult to satisfy these needs through conventional retailing. The towns were too small and far away to support more than the most basic shops. Mail order therefore met a very real need and the Sears Roebuck catalogue became a staple item in the homesteads of the Mid West.

The idea of mail order spread quickly to other countries in a variety of forms. In Britain it played a significant role in retailing throughout the twentieth century. Generally mail order has taken up to ten per cent of retail sales, mainly in clothing and textiles. In Britain the share has been less, and less than other European countries, notably Germany. This is possibly because Britain has been a more urban society than most. Mail order tends to be strongest in rural areas where it is difficult to reach good comparison shopping, and only a small minority of the British population falls into that category. Mail order has many of the same advantages as Internet shopping and it could well be that as a result it feels competition from the Internet more than conventional shops.

Other forms of home shopping include direct mail and newspaper and telephone selling. Direct mail involves sending an offer through the post to customers. Often it is combined with newspaper advertising. Examples are book clubs, wine clubs and specialist areas. Some selling takes place directly through advertising in the newspapers and magazines, and sometimes using coupons. These methods have a similarity to mail order, the main difference being that they avoid the use of agents. Mail order in Britain has tended to involve door-to-door calling by agents who leave catalogues. These other forms of direct selling avoid that cost.

Finally we need to mention the telephone. Ordering goods by telephone, whether from a shop or a warehouse, is an easy and simple method of home shopping, and in many ways is a competitor or precursor of the Internet. Although the telephone had existed for a century it took a long time before its use was widespread. It was not until the last quarter of the century that the telephone became a near universal household convenience. As a result the widespread use of the telephone for domestic business is quite recent. The late 1990s saw an explosion in the development of call centres which exploited breakthroughs in telecommunication technology to offer a service which could cope with an ever growing domestic demand. It has been suggested that the Internet will make this investment in call centres redundant, but that remains to be seen. Even for households with access to the Internet via a home computer, there could still be attractions in picking up the phone and talking to a human being.

The advantages of the Internet

The Internet has three main advantages: convenience, choice and cost. Of these the first is one it shares with the other forms of home shopping. Shopping via the Internet avoids the cost and time of making a journey, the chore of leaving the house, travelling along congested roads, sometimes paying for the privilege of parking, being jostled on crowded pavements and finally standing in a queue at the checkouts. The convenience of being able to avoid this is very appealing, particularly for the weekly chore of food shopping.

It may seem strange that as food shopping has become easier the shopper appears to find it less enjoyable. There are two reasons for this. The first relates to standardisation and branding. Food has increasingly been

Table 12.1 Most popular purchases on the Internet

1	Books	52.9%
2	Travel	29.4%
3	Music	20.6%
4	Clothes	14.7%
5	Electrical Goods	8.8%

Source: Greig & Stephenson, *Retail Week*, November 1999,
supplied by Insignia Richard Ellis

packaged in standard form and branding has become more common on the supermarket shelf. As a result some of the pleasure of intimacy with the product has been lost. In the old days there was personal contact, too, with the shop staff. Some of this survives with delicatessen counters and shopping in street markets and specialist food shops but overall supermarket shopping although more efficient and quicker than food shopping used to be is often now felt to be a burden.

The second reason for considering food shopping a chore is the change in social customs. Increasingly households consist mainly of a couple who are both working which means that the visit to the supermarket has to encroach on leisure time and evenings and weekends. A form of shopping which can avoid this will therefore be seen as attractive to many people. Despite the yearning to escape the weekly trip to the supermarket the indications are that this is likely to be a fairly minor part of Internet shopping in the future. Figures on Internet usage (see Table 12.1) suggest that specialist items better exploit its advantages.

The second main advantage of the Internet is choice, in fact it is by far the biggest advantage. Unlike mail order or the other forms of home shopping, the Internet offers access to an enormous range of information. Someone seeking to buy a specialist item can compare offers not only in one country but from anywhere in the world. Whereas the mail order catalogue customer is limited to what is available in the catalogue, there is no limit to the choice to which the Internet user has access. The advantage of choice also applies to the comparison with conventional shopping. No shopping centre or any conglomeration of shops can hope to match the range available on the Internet.

It follows that abundant choice gives particular advantage to the shopper seeking more specialist and expensive goods. It is here that a wide choice really counts. As the table shows, the most obvious example of this is books. There are so many different titles, hundreds of thousands, that even the biggest book store can stock only a fraction. A service which enables a specific title to be tracked down and its availability and price checked has enormous advantages to the book buyer. Of course this choice is available not only to the home-based Internet user but also to the shopper who prefers to phone or walk into a book shop and ask the shop to undertake the search.

Table 12.2 Cheaper on the Net: the difference between
high street and Internet prices

Books (paperback)	−22%
Vacuum cleaners	−15%
Widescreen TVs	−14%
Videos	−14%
Fridge-Freezers	−12%
Washer-dryers	−11%
Microwaves	−10%
Compact discs	−8%
Toys	0%
Cassette tapes	10%

Source: Barclays Capital, cited by Dr Sushil Wadhani of the
MPC, supplied by Insignia Richard Ellis.

Note
Figures for December 1999. A minus number indicates a lower
Internet price.

The main disadvantage of this second approach is price. The bookshop will
charge the full retail price whereas the Internet supplier is likely to sell at a
discount, the third of the advantages of Internet shopping.

Before we consider the price advantage it is worth looking at some of the
other fields where Internet choice is particularly telling. Table 12.1 mentions
travel and music after books as the most popular on-line purchases. Food is
not included although survey evidence suggests that a third of interviewees
were interested in Internet food shopping. So far desire has not been
converted into reality.

Other areas where choice is important are fields such as hi-fi and
computer peripherals where the goods are all branded. Branding is
particularly suitable for Internet shopping because, as we saw in chapter 8,
branding commonly means that the decision to buy is made without
needing to inspect the goods.

The third major advantage of the Internet lies in price. In general, goods
sold on-line are cheaper than those sold through shops. This is because the
Internet provider or e-tailer does not have the expense of shop premises or
sales staff, which as we saw in the example in chapter 7 may amount to some
28 per cent of sales income. The e-tailer of course has costs too, but in
general they are less than the conventional bricks and mortar retailer. Both
types need to buy the goods and distribute them, which as is shown in the
next section is a problem for the e-tailer. Both he and the conventional
retailer need to spend on advertising but from the point of view of fixed
overheads, prices in conventional shops are likely to be higher.

It is uncertain at the time of writing how big the Internet discount will be.
Figures published in December 1999 and shown in Table 12.2 suggest that
discounts are usually in the 10–15 per cent range. This is a useful but not

overwhelming advantage over conventional shopping. Whether Internet shopping will be able to offer deeper discounts in future depends on whether e-tailers are able to build up economies of scale and how they solve the problem of delivery which is the subject of the next section.

The problems of delivery

The main disadvantage of Internet shopping, as with other forms of home shopping, is that the shopper does not have instant access to the goods. To be able to touch and see the goods, discuss their qualities with a knowledgeable assistant and walk out of the shop with them is an enormous plus and the main reason why up till now home shopping has remained a fairly minor element in retailing. With home shopping there is an inevitable delay between ordering and receiving the goods.

The problem of how to deliver goods to the home is the biggest difficulty facing the Internet and it is a challenge which it needs to overcome if it is to develop into a major force in retailing. To do so it will be necessary to overcome a long running trend which has been going in the opposite direction. In the first half of the twentieth century, when domestic servants were common, home delivery was widespread. Delivery boys called from the grocer, butcher and similar trades and took orders which were then delivered later in the same day. They had special bicycles built with a goods carrying section on the handlebars.

Gradually this system declined with the disappearance of servants but still in the 1950s and 1960s home delivery by bakers and greengrocers was widespread. There was someone in the house when the van arrived and there were obvious advantages in having bulky staples such as bread and potatoes delivered to the door. The arrival of supermarkets and widespread car ownership meant that as an alternative it was possible to buy all the weekly food shopping on a single trip. The choice was wider, the goods were fresher, and above all they were cheaper. They were cheaper because the super-market enjoyed economies of scale and the cost of transporting the goods from shop to home was switched from the retailer to the shopper. As a result home delivery declined further and all that remained was the milkman. The milkman is a particularly British institution, home delivery of milk being far less common in Europe and America. It survives because its deliveries take place very early and in many households it is possible to take the milk bottle from the doorstep to the 'fridge before leaving for work. Nevertheless milk is increasingly being bought in supermarkets and the home delivery of milk is in decline.

Cost is one of the basic problems of home delivery. However it is achieved, carrying goods to an individual householder's door is labour intensive and expensive. The cost of it goes a long way to neutralise the benefit which home shopping has in avoiding the expense of running a shop. The biggest cost of home delivery is labour, and the comparison is between the labour of the shopper in visiting the shops and the labour of the delivery man. People tend not to put a financial value on their own time, and although overall the Internet system may be more efficient than having lots of individual shoppers all travelling about, the former has a cost while the latter appears not to.

To the cost of delivery should also be added the cost of picking the goods from the shelves. Although automated warehouses are developing fast, picking is still a labour intensive activity. Again this is a task which the shopper performs free and it is a cost which falls on Internet shopping and not on the conventional supermarket. Certainly the cost burden of both picking and home delivery falls heaviest on food shopping. These problems are less significant for high value to weight items like books, music and electronic gadgets as mentioned above. But for food they are a major factor. Supermarkets both in Britain and America which offer home delivery usually make a charge. At the time of writing the going rate is £5 per delivery, but it is clear that this is below the real cost. There is certainly a market for home delivery of food by the cash-rich time-poor members of society but the cost will be high. There is likely to be a substantial premium rather than a discount for Internet food shopping.

Cost is not the only problem of home delivery. There is also the physical difficulty of actually delivering goods to the home. As anyone knows who has attempted to collect money for charity door-to-door, a ring on the door bell produces no answer on many occasions, even in the evening. A typical family will have both adult members out working during the day when delivery might normally take place. The result is often frustration for the delivery of anything which is too big to be posted through a letterbox.

Various ways have been put forward of getting round this problem. One possibility is to install a delivery box in the porch or near the front door which would function like a giant letterbox. If this were able to accommodate a food delivery it would need to be refrigerated. Other alternatives put forward are to use local post offices, stations or garages as delivery points. The goods would be placed in special lockers to await collection. The disadvantage of this is that the shopper would still have to make a trip to collect the goods, albeit one which is shorter than the trip to the supermarket. This eliminates the main advantage of home shopping, which is to avoid making a trip. For many it is the journey itself rather than its length which they find irksome.

It is difficult to see how the problem of home delivery will be solved for food shopping without a substantial cost in one form or another. For other types of product the problem is less acute. There are existing postal and parcel delivery services but the difficulty of finding someone at home to accept the goods still remains to be solved. Inevitably there is a delay between order and delivery, uncertainty when the goods will arrive and the problem of correcting any mistake which is made. The problems of home delivery remain to be solved.

Clicks and mortar

In locational terms the effect of the Internet on retail should be to accelerate the existing trend which has been described throughout this book. In summary these are continued gradual contraction of retailing in many town centres and its replacement by services, particularly those offering places to eat and drink. Attractive town centres will prosper while decline will be most marked in the remainder.

It is likely that in addition to diverting some sales from traditional shops, the Internet could change the way conventional shops operate. The major multiples have already indicated that they are not prepared to watch while the new breed of dot-com e-tailers steal their business. They have set up their own websites and they have the advantage of offering the comfort of a well known name and a proven supply system. This offers the prospect of greater flexibility. Having used the Internet to select an item offered by an existing bricks and mortar retailer, for example, the shopper could then visit the shop to inspect it, discuss it with sales staff or see it demonstrated and then arrange to have it delivered either to the shop or direct to home. It is possible to think of a number of permutations of how this combination of the Internet and existing shop property could operate. In combining the clicks of the Internet with the mortar of the physical shop it would be possible to exploit the advantages of both systems. The shopper could use the Internet to search and get the best terms and the shop for the fulfilment of the purchase.

The advantage of this is that the shop no longer suffers from being able to carry only a limited stock of goods. The shop might itself carry an Internet screen offering a wide range of items which it could guarantee to produce from a nearby storage site in a short period of time. This is similar to the system already offered by catalogue showrooms such as Argos. The Internet then becomes a powerful addition to the catalogue and the shop turns more into a showroom.

Many commentators expect the Internet to have its biggest impact on businesses which sell to other businesses rather than the general public. It is easy to see how for the buyer of business goods and services the unequalled ability of the Internet to offer complete information on what is on offer will be a boon. It is likely to have a major effect on marketing.

It is difficult to predict what impact the Internet will have on business location. Important though it undoubtedly is, the Internet is a tool like electricity or the telephone which will increasingly be exploited in those areas where it offers benefit. On the evidence of the other great innovations such as the two just mentioned, the Internet could, as with shopping, accelerate changes which are already happening. It is less a case of winner and losers than one of users and losers.

The change most likely to be accelerated is the move away from pen-pushing and data processing towards greater face-to-face contact. This means more meetings, more travelling and more emphasis on selling and understanding whatever market the firm is serving. In locational terms the effect should accelerate the existing trend away from the pattern of nine to five working in centrally located offices towards more dispersed locations and more diverse forms of working, such as working from home. This in turn frees businesses and their workers increasingly to locate where they please which will increase pressure on rural and coastal areas. In conclusion the locational effect of the Internet on business is unlikely to be revolutionary in the short term. As with other great inventions the change in location which it brings about will be gradual and difficult to disentangle from other causal factors. As far as it is possible to discern, the effect flows in the same direction as trends which are already in existence. This makes it even harder to separate out the effect of the Internet.

Regarding the effect on shopping, it is important to remember that the two biggest categories of goods bought in shops are food and clothing, neither of which are particularly suited to exploit the advantages of the Internet. In the case of food, problems of delivery are likely to remain a major inhibitor. For clothing most people will continue to prefer to buy in a shop where they can examine and try on the clothes before buying them. The Internet seems much better suited to services where the problem of delivery does not arise. It is ideal for selecting holidays, booking theatre tickets and restaurant tables and similar activities. Another example might be buying new tyres for a car where a garage website might offer details not only on price and availability but also on times when the car could be booked in to have the tyres fitted.

The demand for services, including those delivered to the door will certainly grow. The cash-rich time-poor consumer of the future will want gratification fast, including delivery to the door within an hour if possible. An example of just such a service is Urban Fetch which operates in New York and has opened in London. They will deliver (according to Joanna Coles writing in *The Times*, 25 October 1999): '24 hour supplies of Hershey bars, steaming cappuccinos, the new Keanu Reeves video or Melissa Etheridge CD, tubs of Ben and Jerry's, the latest edition of *Talk* and every other magazine, newspaper or book published post Caxton.'

No doubt the charge for such services will be substantial but there appears already to be a demand among the rich flat-dwellers of Manhattan and the proportion of the population able and wanting to join them is likely to spread. We already live in a service economy and the importance of services relative to objects seems certain to grow. The Internet speeds this change along. In summary, then, the Internet should have a fairly marginal effect on the way we shop, at any rate as far as it is prudent to look. The effect in other areas is likely to be profound but the demand for property will remain and with it the principles of location which underpin it.

Bibliography

1 Introduction

Garreau, J. (1991), *Edge City*, Doubleday, New York.

2 Location and politics

Barrett, S. and Healey, P. (eds) (1985), *Land Policy: Problems and Alternatives*, Gower, Aldershot.

Davies, R.L. (ed.) (1979), *Retail Planning in the European Community*, Saxon House, Farnborough.

Davies, R.L. (1984), *Retail and Commercial Planning*, Croom Helm, London.

Dawson, J.A. and Lord, J.D. (1985), *Shopping Centre Development: Politics and Prospects*, Croom Helm, London.

Healey, P. (1983), *Local Plans in British Land Use Planning*, Pergamon, Oxford.

McFadyen, E. (ed.) (1987), *The Changing Face of British Retailing*, Newman Books, London.

Ministry of Housing and Local Government (1962a), *The Green Belts*, HMSO, London.

Ministry of Housing and Local Government (1962b), *Town Centres: Approach to Renewal*, HMSO, London.

Ministry of Housing and Local Government (1965), *Parking in Town Centres*, Planning Bulletin No. 7, HMSO, London.

Schiller, R.K. (1975), The impact of new shopping schemes on shops in historic streets, *The Planner*, December.

3 The importance of technology

Dunphy, F. *et al.* (1997), *Moving Beyond Gridlock: Traffic and Development*, The Urban Land Institute (ULI), Washington DC.

Haas-Klau, C. *et al.* (1998), *Accessibility, Walking and Linked Trips*, a report for NRPF and DETR, 6 Copperfield Street, London SE1 0EP.

IGD (1993), *Retail Logistics*, Institute of Grocery Distribution, Watford.

Starkie, D. (1982), *The Motorway Age*, Pergamon Press, Oxford.

4 The importance of value

Edwards, M. and Lovall, D. (1980), *Understanding Urban Land Values: A Review*, SSRC Inner Cities Working Party, ISBN 0-86226-0191.

Henneberry, J. (1995), Development cycles in British cities, in S. Hardy, M. Hebbert, and B. Malton (eds), *Region – Building*, Proceedings of the Regional Studies Association Annual Conference, pp. 44–52.

Henneberry, J. (1996), *Property Market Structure and Behaviour: The Interaction of Use and Investment Sectors and its Impact on Urban and Regional Development*, Proceedings of Cutting Edge Property Research Conference, RICS.

Henneberry, J. (1997), *Rent–Yield Asynchrony and the Property Cycle*, Proceedings of Cutting Edge Property Research Conference, RICS.

CB Hillier Parker (1994), *Hillier Parker Rent Index Digest*, CB Hillier Parker, London, and following quarterly publications.

CB Hillier Parker (1997), *Central London Shops Survey*, CB Hillier Parker, London.

Howes, C.K. (1980), *Value Maps: Aspects of Land and Property Values*, Geo Abstracts, Norwich.

Marx, K. (1867–95), *Das Kapital*.

Mill, J.S. (1848), *Principles of Political Economy*.

Ricardo, D. (1817), *Principles of Political Economy and Taxation*.

RICS (1994a), *Economic Cycles and Property Cycles*, Department of Land Economy, University of Aberdeen and Investment Property Databank, RICS, London.

RICS (1994b), *The Mallinson Report: Report to the President's Working Party on Commercial Property Valuations*, RICS, London.

Urbed, B. *et al.* (1994), *Vital and Viable Town Centres: Meeting the Challenge*, DoE, HMSO, London.

5 A little theory

Applebaum, W. (1965), Can store location research be a science?, *Economic Geography*, 41, 234–7.

Berry, B. (1963), *Commercial Structure and Commercial Blight*, Research Paper No. 85, Dept. of Geography, University of Chicago, Chicago.

Berry, B. (1967), *Geography of Market Centres and Retail Distribution*, Prentice Hall, Englewood Cliffs, USA.

Black, J. (1966), Some retail sales models, Paper to Urban Studies Conference, Oxford.

Brown, S. (1993), Postmodernism and central Place Theory: deconstruction/reconstruction, Proceedings of the 7th International Conference on Research in the Distributive Trades, ed. S. Bent and L. Sparks.

Clarke, C. (1951), Urban population densities, *Journal of the Royal Statistical Society*, Series A, 114, 490–6.

Christaller, W. (1933), *Central Places in Southern Germany*, Prentice Hall, Englewood Cliffs, USA (translated by C. Baskin, 1966).

Cohen, S.B. and Lewis, G.K. (1967), Form and function in the geography of retailing, *Economic Geography*, 53, 1–42.

Davies, R.L. (1976), *Marketing Geography*, Retailing and Planning Association (RPA), Corbridge, Northumberland.

Davies, R.L. and Rogers, D.S. (eds) (1984), *Store Location and Store Assessment Research*, Wiley, Chichester and New York

Lomas, G.M. and the Research Group of the West Midlands RTPI (1967), *Predicting Shopping Requirements*, West Midlands Branch of the Town Planning Institute.

Hagerstrand, T. (1957), Migration and area: survey of a sample of Swedish migration fields and hypothetical considerations on their genesis, *Lund Studies Series*, B13.

Haggett, P. (1965), *Locational Analysis in Human Geography*, Edward Arnold, London.

Hammer, C. and Ikle, F.C. (1957), Intercity telephone and airline traffic related to distance and the 'propensity to interact', *Sociometry*, vol. 20.

Huff, D.L. (1963), A probabilistic analysis of shopping centre trade areas, *Land Economics*, 39, 81–90.

Jones, C.S. (1969), *Regional Shopping Centres*, Business Books, London.

Kornblau, C. (ed.) (1968), *Guide to Store Location Research*, Addison-Wesley, Reading, Mass.

Lakshmanan, T.R. and Hansen, W.G. (1965), A retail market potential model, *American Institute of Planners Journal*, vol. xxxi.

Lösch, A. (1940), *The Economics of Location*, Yale University Press, New Haven (translated by W.H. Woglam and W.F. Stolper, 1954).

Lowry, I.S. (1964), *Model of Metropolis*, Rand Memo RM4035 RC20.

Ministry of Housing and Local Government (1967), *A Method of Estimating Retail Floor space Requirements in Small and Medium Sized Towns Outside Conurbations*, HMSO, London.

Nelson, R.L. (1958), *The Selection of Retail Locations*, McGraw Hill, New York.

Oxford Institute of Retail Management (OXIRM) and Jones Lang Wootton (JLW) (1966), *Retail Planning Policies, Their Impact on European Retail Property Markets*, Oxford Institute of Retail Management and Jones Lang Wootton, JLW, London.

Reilly, W.J. (1929), Methods for the study of retail relationships, University *of Texas Bulletin*, No. 2944.

Schiller, R.K. (1971), 'Model of a Marketing Area', unpublished Ph.D. thesis, Dept of Geography, University of Reading.

Smith, Larry and Gruen, V. (1959), *Shopping Towns*, Reinhold, Orlando, Florida.

South Beds Sub-Regional Study (1967), *Technical Committee Shopping Report*, Bedfordshire County Planning Dept.

Stewart, J.Q. (1948), Demographic gravitation: evidence and applications, *Sociometry*, vol. 11.

Thünen, J.H. von (1875), *Der Isolierte Staat in Beziehung auf Landwirtschaft und Nationalökonomie*, Hamburg.

Urbed, B. *et al.* (1994), *Vital and Viable Town Centres: Meeting the Challenge*, DoE, HMSO, London.

Weber, A. (1909), *Über den Standort der Industrien*, Tubingen.

Wilson, A.G. (1967), A statistical theory of spatial distribution models, *Transportation Research*, Nov.

Wrigley, N. (ed.) (1988), *Store Choice, Store Location and Market Analysis*, Routledge, London.

Zipf, G.K. (1949), *Human Behaviour and the Principles of Least Effort*, Addison Wesley Press.

6 Hierarchy

Applebaum, W. (1957), Towards a better understanding of store site evaluations and rentals, in American Marketing Association, *The Frontiers of Marketing Thought and Science*, New York.

Auerbach, F. (1913), Das Gesetz der bevolkeriengskonzentration, *Petermann's Mitteilungen*, 59, 74–6.

Berry, B.L. (1961), City size distributions and economic development, *Economic Development and Cultural Change*, 9, 573–88.

Berry, B.L. (1967), *Geography of Market Centres and Retail Distribution*, Prentice Hall, New York.

British Council of Shopping Centres (1998), *The Shopping Centre Industry, Its Importance to the UK Economy*, BCSC, London.

Carruthers, W.I. (1957), A classification of service centres in England and Wales, *The Geographical Journal*, 123, 371–85.

Davies, R.L. (1976), *Marketing Geography*, Retail and Planning Association (RPA), Corbridge, Northumberland.

Distributive Trades EDC (1971), *The Future Pattern of Shopping*, HMSO, London.

Distributive Trades EDC (1988), *The Future of the High Street*, HMSO, London.

DTZ (1999), Factory outlet shopping centres, *Retail Week*, 29 January 1999.

Haggett, P. (1965), *Locational Analysis in Human Geography*, Edward Arnold, London.

Hillier Parker (1990), *Retail Warehouse Rent Index*, No. 2, September 1990, Hillier Parker, London.

Hillier Parker (1995), *British Shopping Centre Development Master List*, Hillier Parker, London.

Hillier Parker (1996), *Shopping Centres of Great Britain – a National Survey of Retailer Representation by Traditional Location*, Hillier Parker, London.

Nelson, R.L. (1958), *The Selection of Retail Locations*, Dodge, Chicago.

Reynolds, J. and Schiller, R.K. (1992), A new classification of shopping centres in Great Britain using multiple branch numbers, *Journal of Property Research*, 9, 122–60.

Schiller, R.K. (1971), Location trends in specialist services, *Regional Studies*, 1–10.

Schiller, R.K. and Jarrett, A. (1985), A ranking of shopping centres using multiple branch numbers, *Land Development Studies*, 2, 53–100.

Smailes, A.E. (1944), The urban hierarchy in England and Wales, *Geography*, 29, 21–55.

Smith, R.D.P. (1968), The Changing urban hierarchy, *Regional Studies*, 4, 85–96.

Stewart, C.T. (1958), The size and spacing of cities, *Geographical Review*, 48, 222–45.

Teale, M. (1997), Tracking the consumer, *Estates Gazette*, issue 9709, March 1997.

Thorpe, D. (1968), The main shopping centres of Great Britain in 1961, the location and structural characteristics, *Urban Studies*, 5, 165–206.

7 The desire to cluster

Garrison, W. *et al.* (1959), *Studies of Highway Development and Geographic Change*, Seattle.

Hillier Parker (1997), *Central London Shops Survey*, Hillier Parker, London.

Kirkup, M. and Rafiq, M. (1994a), Tenancy development in new shopping centres: implications for retailers and developers, *International Review of Retail, Distribution and consumer Research*, 4(3), 17–32.

Kirkup, M. and Rafiq, M. (1994b), Managing tenant mix in new shopping centres, *International Journal of Retail and Distribution Management*, 22(6), 29–37.

Schiller, R.K. (1996), Town centre winners and losers, *Estates Gazette*, 13 July, London.

8 The desire to disperse

Guy, C. (1994), *The Retail Development Process: Location, Property and Planning*, Routledge, London.

Hall, P. and Markensen, A. (eds) (1985) *Silicon Landscape*, Allen and Unwin, Boston.

CB Hillier Parker and Savell Bird Axon (1998), *The Impact of Large Foodstores on Market Towns and District Centres*, DETR, HMSO, London.

Hillier Parker (1995), *Shopping Centres of Great Britain: a National Survey of Retail Representation by Trading Location*, Hillier Parker, London.

Hillier Parker (1996), *Fashion Designers in Central London*, Hillier Parker, London.

Hillier Parker (1997), *Fashion Designer Stores Expansion in London and York*, Hillier Parker, London.

Schiller, R. (1986) Retail decentralisation: the coming of the third wave, *Estates Gazette*, 16 August.

Saxenian, A. (1983), The genesis of Silicon Valley, *Built Environment*, 9, 7–17.

Urbed, B. *et al.* (1994), *Vital and Viable Town Centres: Meeting the Challenge*, DoE, HMSO, London.

9 Office location

Aksoy, A. and Marshall, N. (1991), The changing corporate head office and its spatial implications, *Regional Studies*, 26(2), 149–62.

Daniels, P.W. (1969), Office decentralisation from London: policy and practice, *Regional Studies*, 3, 171–8.

Diamond, D.R. (1991), The City, the 'Big Bang' and Office Development, in K. Hoggart and D.R. Green (eds), *London: A New Metropolitan Geography*.

Evans, A.W. (1985), *Urban Economics, an Introduction*, Basil Blackwell, Oxford.

Garreau, J. (1991), *Edge City*, Doubleday, New York.

Goddard, J.B. (1968), *Office Linkages and Location*, Pergamon Press, Oxford.

Goddard, J.B. and Pye, R. (1976), Telecommunications and office location, *Regional Studies*, 11, 19–30.

Gottman, J. (1966), Why the skyscraper?, *The Geographical Review*, New York, vol. lvi.

Hillier Parker, (1990), *European Atlas of Office Rent Contours*, Hillier Parker, London.

Hinds, D.S. and Corgel, J.B. (1984), *Understanding the Effect of Transportation on Office Location Perspective*, SIREF, USA.

Manners, G. (1986), Decentralizing London, 1945–1975, in H. Clout and P. Woods (eds), *London: Problems of Change*.

Moss, M. (1987), Telecommunications and international financial centres, in J.F. Brotchie, P. Hall and P.W. Newton (eds), *The Spatial Impact of Technological Change.*

Rosen, K. (1996), The real estate recovery and the real estate cycle, the US experience, paper at the Cambridge–Wharton Conference, Dept of Land Economy, Cambridge.

St Quintin (1994), *The City of London to the Year 2000 and Beyond, Prospects for Office Demand*, St Quintin, for Corporation of London *et al.*

Schiller, R.K. (1988), Office decentralization, lessons from America, *Estates Gazette*, 9 April.

10 Industrial location

Chapman, K. and Walker, D. (1987), *Industrial Location*, Basil Blackwell, Oxford.

Henneberry, J. (1987), *British Science Parks and High Technology Developments: Progress and Change, 1983–1986*, PAVIC, Sheffield City Polytechnic, Sheffield.

Hillier Parker (1980), *Industrial Rent Contours*, CB Hillier Parker, London.

Hillier Parker (1982), *Industrial Contour Map Rents and Yields*, CB Hillier Parker, London.

Hillier Parker (1987), *Hi-tech Users*, CB Hillier Parker, London.

Hillier Parker (1994 and 1999), *Hillier Parker Rent Index Digest*, CB Hillier Parker, London.

Schiller, R. (1996), High bay warehouses: a new type of investment property, *Property Review*, April 1996.

Segal Quince Wickstead (1985), *The Cambridge Phenomenon: The Growth of High Technology Industry in a University Town*, Segal Quince Wickstead, Cambridge.

Weber, A. (1929), *Theory of the Location of Industries* (translated by C.J. Friedrich), University of Chicago, Chicago.

11 People and houses

Aldridge, Eastman & Waltch *et al.* (1990), *Land Value Study – Boston 1850–1989*, Aldridge, Eastman & Waltch, Federal Reserve Bank of Boston, and Grantham, Mayo, Van-Otterloo & Co, Boston, USA.

Alonso, W. (1964), *Location and Land Use*, Harvard University Press, Cambridge, MA.

Bibby, P. and Shepherd, J. (1996), *Urbanisation in England: Projections 1991–2016*, DoE, HMSO, London.

Cabinet Office (1999), Sharing the nation's prosperity – variation in economic and social conditions across the UK, a report to the Prime Minister by the Cabinet Office, London, December 1999.

Champion, A. (ed.) (1989), *Counterurbanisation, the Changing Pace and Nature of Population Deconcentration*, Edward Arnold, London.

Champion, A. and Dorling, D. (1994), Population change for Britain's functional regions 1951–91, *Population Trends*, 77, OPCS, HMSO, London.

Daly, (1971), Characteristics of 12 clusters of wards in Greater London, *Research Report No. 13*, Department of Planning and Transportation, Greater London Council.

DoE (1988), *Urban Land Markets in the United Kingdom*, DoE, HMSO, London.

DoE, *Projections of Households in England to 2016, 1992-based Estimates*, DoE, HMSO, London.

Edwards, M. and Lovatt, D. (1980), *Understanding Urban Land Values: a Review, the Inner City in Context*, vol. 1, Bartlett, UCL, London.

Evans, A. (1973), *The Economics of Residential Location*, Macmillan, London.

Evans, A. (1987), *House Prices and Land Prices in the South-East – A Review*, the Housebuilders Federation, London.

Evans, A. (1996), The land market and government intervention: a review of the empirical evidence, *Discussion Paper No. 117 in Urban and Regional Economics*, FIRS, University of Reading.

Gerald Eve *et al.* (1992), *The Relationship between House Prices and Land Supply*, Gerald Eve and Department of Land Economy, University of Cambridge for DoE, HMSO, London.

Government Actuary's Department (1999), *National Population Projections – 1996 Based*, HMSO, London.

Haas-Klau, C. (1998), *Accessibility, Walking and Linked Trips*, National Retail Planning Forum, London.

Hall, P. *et al.* (1973), *The Containment of Urban England*, George Allan and Unwin.

Lloyd, G. and Rowan-Robinson, J. (1987), *The Social Costs of Land Development*, A Report to the Scottish Development Agency, Department of Land Economy, University of Aberdeen.

Mean, G. (1997), Regional house prices and the ripple effect: a new interpretation, *Discussion Paper No. 126*, Department of Economics, University of Reading.

Monk, S. (1999), The use of price in planning for housing: a literature review, *Discussion Paper No. 105*, Property Research Unit, Department of Land Economy, University of Cambridge.

The Times (2000), New York Upper East Side, *The Times*, 7 January.

OPCS (1997), *1991 Census, Key Statistics for Urban and Rural Areas in Great Britain*, ONS, HMSO, London.

Shucksmith, M. *et al.* (1995), *A Classification of Rural Housing Markets in England*, DoE, HMSO, London.

12 Home shopping and the Internet

DTZ Research (1999), *The Effects of Electronic Business on Real Estate*, DTZ Debenham Thorpe, London.

Retail Week (2000), E-Commerce – opportunities for retailers, *Retail Week Supplement*, 28 April.

Index